Research in Collegiate Mathematics Education. V

Issues in Mathematics Education

Volume 12

Research in Collegiate Mathematics Education. V

Annie Selden
Ed Dubinsky
Guershon Harel
Fernando Hitt
Editors

Cathy Kessel, *Managing Editor*
Michael Keynes, *Assistant Managing Editor*

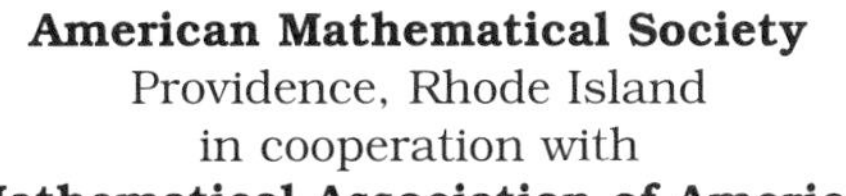
American Mathematical Society
Providence, Rhode Island
in cooperation with
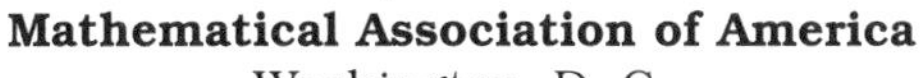
Mathematical Association of America
Washington, D. C.

2000 *Mathematics Subject Classification*. Primary 00–XX, 97–XX.

ISBN 0-8218-3302-2
ISSN 1047-398X

∞ The paper used in this book is acid-free and falls within the guidelines
established to ensure permanence and durability.
Visit the AMS home page at `http://www.ams.org/`

10 9 8 7 6 5 4 3 2 1 08 07 06 05 04 03

Contents

Preface vii
Anne Selden, Ed Dubinsky, Guershon Harel, Fernando Hitt

First-year Undergraduates' Difficulties in Working with Different Uses of Variable 1
María Trigueros and Sonia Ursini

Cooperative Learning in Calculus Reform: What Have We Learned? 30
Abbe Herzig and David T. Kung

Calculus Reform and Traditional Students' Use of Calculus in an Engineering Mechanics Course 56
Cheryl Roddick

Primary Intuitions and Instruction: The Case of Actual Infinity 79
Pessia Tsamir

Student Perfomance and Attitudes in Courses Based on APOS Theory and the ACE Teaching Cycle 97
Kirk Weller, Julie M. Clark, Ed Dubinsky, Sergio Loch, Michael A. McDonald, and Robert R. Merkovsky

Models and Theories of Mathematical Understanding: Comparing Pirie and Kieren's Model of the Growth of Mathematical Understanding and APOS Theory 132
David E. Meel

The Nature of Learning in Interactive Technological Environments: A Proposal for a Research Agenda Based on Grounded Theory 182
Jack Bookman and David Malone

Preface

Welcome to the fifth volume of Research in Collegiate Mathematics Education *(RCME V)*. This and the four previous volumes serve purposes similar to those of a journal. Each presents readers with peer-reviewed research on questions regarding the teaching and learning of collegiate mathematics.

There is now a growing international community of mathematics education researchers working at the post-secondary level, asking pertinent questions, gathering and analyzing data, and reflecting upon a variety of pedagogical issues. Like mathematics education research at other levels, potential topics and viewpoints are extremely broad and varied. One can take a cognitive point of view, asking how individual learners come to understand specific mathematical concepts such as limit, or one can take a more socio-cultural stance, examining classroom situations to uncover ways, such as the structuring of whole class and small group discussions, that can facilitate learning. Depending on the question posed, one might decide to conduct a qualitative or quantitative study, or a combination of both. Researchers often begin with a theoretical framework—a "lens" through which to analyze their data—and build on the work of others. However, sometimes when little is known, an exploratory study may be a way of getting a "feel for the territory."

Collegiate mathematics education research is a relatively young field, but interest in it and its results is growing. Some things are known, but much remains to be uncovered. The editors of *RCME* are pleased to be part of this development as they aid researchers in bringing such studies before a combined audience of colleagues, both fellow researchers and mathematicians with an interest in teaching and learning.

Overview of *RCME V*

The volume begins with a study from Mexico of students' understandings of variable, a concept that pervades mathematics. It is followed by two studies dealing with aspects of calculus reform. The first of these notes that calculus reform projects have used a variety of interacting innovations to achieve their effects and tries to isolate information on the contribution of cooperative learning, while the second compares reform and traditional students' uses of calculus in a subsequent engineering mechanics course. The next article reports a study of Israeli preservice teachers' intuitions regarding the concept of actual infinity, using Fischbein's ideas on intuition. This is followed by two articles that concern APOS theory. The first summarizes published and unpublished studies regarding students' performance and attitudes in courses based on APOS (Action, Process, Object, Schema) theory and the ACE (activities, class discussion, exercises) teaching cycle. The second compares and contrasts the views of mathematical understanding provided by APOS theory and Pirie and Kieren's model for the growth of mathematical understanding. Finally, there is a consideration of the kinds of research questions that might be asked to gain insights regarding the nature of learning in interactive technological environments. This broad spectrum of research coming from three different countries provides insights, as well as concerns, regarding a broad range of common collegiate mathematics teaching and learning problems.

The Concept of Variable and Calculus Reform

A flexible understanding of algebraic concepts and notation, in particular the concept of variable, is important for making progress in most university mathematics courses. Although variable as general number, as specific unknown, and as indicator of a variable quantity in a functional relationship are all taught at the school level, María Trigueros and Sonia Ursini's analysis suggests that its multifaceted nature is often not made clear to secondary students. Consequently, beginning college students often seem unable to deal with variables flexibly. To test that conjecture, Trigueros and Ursini developed a questionnaire of open-ended items to investigate first-year undergraduates' knowledge of aspects of variable.

Trigueros and Ursini developed a framework for analysis of this knowledge; that is, for each aspect, they developed a description of levels of abstraction at which it might occur. For specific unknown, these range from the ability to recognize and identify something in a problem situation that can be determined to the ability to symbolize unknown quantities in specific situations and write relevant equations about them. After analyzing 164 students' responses to the questionnaire, the researchers concluded that these students' understandings of variable were "very weak." Just 7 of 65 items, involving only very elementary algebra or pattern matching, had more than 90% correct solutions and there were other items on which more than 40% of the students gave incorrect responses.

Trigueros and Ursini also interviewed four students with average questionnaire scores. Their analysis indicates the students had difficulty "discriminating between variable as a specific unknown and variable as a general number." This is typified by the answer "one" in response to the question, "How many values can the letter assume (in the following)? $x+2=2+x$." Students tended to avoid manipulations, preferring to answer questions by inspection. For example, one interviewed student stated that $(x+1)^2 = x^2+2x+1$ would have two answers because it is a quadratic. Interviewed students often gave stock answers based on definitions learned at school, such as "a letter stands for a number" or "a letter represents any number."

The authors conclude that until high school algebra is taught in a way that emphasizes the multi-faceted uses of variable, it may be necessary to design university courses to do this and suggest their framework might be useful for this purpose. Finally, we note that this study was conducted with students at a private Mexican university whose majors ranged from business to political science and whose degree programs required at least six mathematics courses, but beginning university students in the U. S. and elsewhere might well respond similarly.

In their contribution, Abbe Herzig and David T. Kung first note that many calculus reform projects have included a complex interrelated set of innovations that include revised syllabi, use of calculators or computers, emphasis on multiple representations, writing, student projects, and group work. Thus, when examining their effectiveness, it is difficult to isolate which, such as improved student attitudes or increased mathematics course-taking, is due to which innovation. Indeed, it is quite possible that the positive effects are a result of the whole complex of innovations and that the resulting reform cannot be reduced to the sum of its constituent parts. In an attempt to sort out part of this confused situation, Herzig and Kung focused on just one innovation—cooperative learning, and devised a variant of the control group vs. experimental group research paradigm.

They implemented this with ten discussion sections of Calculus I, where the lectures were given by a single professor. Some students had "traditional" discussion sections in which the teaching assistants (TAs) spent most of the time at the board solving problems and answering students' questions with the remainder spent giving quizzes and taking care of administrative details. In the "cooperative" discussion sections, modeled on the Treisman Emerging Scholars Program (ESP), most of the time was spent doing group work with TA-developed worksheets of thought-provoking problems; TAs spent considerably less than half the time at the board. Both kinds of discussion sections were offered in 50-minute or 75-minute sessions, and each format—traditional or cooperative—had sections with one experienced and one novice TA. All TAs participated in a training program before the start of classes, but the TAs doing the "cooperative" sections participated in an additional training program.

Students registered for discussion sections of their choice, mainly based on schedule considerations. To mitigate the resulting non-randomness of the students assigned to each section, Herzig and Kung used the statistical technique of nested linear models and considered only students who had taken calculus in high school, but not previously in college, took all exams and completed the course, and had values for all covariates such as ACT scores and placement tests. The authors developed an attitude survey by adapting four dimensions from the Fennema-Sherman Mathematics Attitudes Scales and adding two of their own. It measured confidence in the ability to learn mathematics, attitudes towards anticipated success in mathematics, beliefs about mathematics' usefulness, motivation to do mathematics, beliefs about mathematics, and views about the helpfulness of learning mathematics with others. They also considered student achievement as measured by scores on the six exams given during the semester.

The authors' main finding is a lack of significant difference between the effects of the two kinds of discussion sections; they speculate on why this might be. The reported success of ESP students, as measured by high achievement scores, might result not only from the greater amount of time spent in group learning, but also could be due to the types of students recruited and the community building activities incorporated. Because the students in this study all had the same type of exams—those given in traditional calculus courses—any problem-solving and higher order reasoning skills developed by cooperative discussion sections may not have been assessed, a possibility noted in other studies of calculus reform where traditional tests have been used to measure achievement. The question of which calculus reform initiatives work for whom, and why, remains open and ripe for more research.

In the next article, Cheryl Roddick reports on her qualitative investigation of the long-term effects of differing approaches to teaching calculus. In particular, she details the differing ways that traditional and Calculus & *Mathematica* students in an engineering mechanics course solved non-routine calculus-related load and moment problems. She conducted task-based interviews with six students enrolled in an engineering mechanics course, whose prerequisites included at least three quarters of calculus and one quarter of physics. All six had been above-average calculus students; three had completed a Calculus & *Mathematica* sequence and the remaining three had completed a traditional calculus sequence. Although other studies have compared Calculus & *Mathematica* students' performances and experiences with those of traditional students (Park & Travers, *RCME II*) or honors students (Meel, *RCME III*), this seems to be the first study of how such students apply their calculus knowledge in subsequent science or engineering courses.

Four interviews were conducted with each student; the first had questions on differentiation and integration similar to those on final exams in first and second quarter calculus and is described elsewhere. In this article, Roddick details these students' thinking (obtained via think-aloud protocols) on specially designed engineering mechanics problems: two non-routine problems in which a load diagram is given and students are asked to produce shear and moment diagrams and two moment extrema problems. She notes that some necessary knowledge for the load problems is that the derivative of the moment function is the shear function, and the derivative of the shear function is the negative load function. There are typically three ways engineering mechanics students solve such problems, two of which are mainly procedural, while the remaining one is conceptual.

Roddick found that the Calculus & *Mathematica* students chose to solve the first (triangular) load problem using their knowledge of slope, area, and the relationship between the shapes of the graphs of a function and its antiderivatives, making use of procedural methods only when these became necessary. The traditional students' methods were more varied. One used a combination of procedural and conceptual methods. Another made several unsuccessful attempts, displaying a misconception, and stopped when he could not come up with equations. The remaining traditional student used a combination of two numerical methods and his understanding of the polynomials generated, as well as how slope is involved in uniformly distributed loads, but this did not help him with this triangular load problem. Similar findings were obtained for the second load problem and the extrema problems.

The Calculus & *Mathematica* students also demonstrated the ability to solve the load problems procedurally, but they preferred to apply their conceptual knowledge of calculus. The traditional students also used some conceptual knowledge of calculus—the general shape of graphs and the idea of slope, but not the integral as the area under a curve. However, they seemed to need specific equations. Roddick interprets these findings using Salomon and Globerson's idea of *mindful abstraction*, in which deliberate attention is paid to decontextualizing, and in terms of the Calculus & *Mathematica* students active learning of calculus with a conceptual focus so as to facilitate the use of knowledge in new situations (i.e., transfer).

Should We Be Appealing to Students' Intuitions?

Pessia Tsamir studied Israeli preservice secondary teachers' tendency to compare the size of infinite sets using the following criteria: inclusion, single infinity (i.e., all infinite sets have the same cardinality), and one-to-one correspondence. She did so for students who had taken a year-long Cantorian set theory course and others who had not. She framed her study using Fischbein's perspective on intuition as "cognition characterized by self-evidence, immediacy, stability, and coerciveness," as well as his distinction between primary and secondary intuitions. *Primary intuitions* develop without systematic instruction as a result of personal experience and are "accompanied by a powerful feeling of certainty." *Secondary intuitions* are new logically based interpretations that develop only under special circumstances (such as teaching) but often cannot compete with primary intuitions regarding the same notion. Thus, students may hold several contradictory views simultaneously.

A number of teaching interventions, some more successful than others, have been tried in order to promote secondary students' psychological acceptance of the one-to-one correspondence criterion for equal cardinality of infinite sets. However, in Israeli colleges of education, year-long set theory courses are taught via traditional lecturing. In these, the one-to-one correspondence criterion is presented as the method for determining cardinality, often with total disregard for preservice students' pre-existing primary intuitions regarding infinity. Tsamir's study focused on the impact of such courses on students' cardinality criteria for infinite sets. She developed a questionnaire to collect information on students' views. Students were first given the questionnaire, then given examples of correct solutions to cardinality problems, and finally the initial questionnaire was readministered to see whether consideration of correct solutions would influence students' responses. In addition, ten students who had taken a set theory course and ten who had not, but all of whom gave insufficient justifications on the questionnaire, were interviewed.

In response to the first administration of the questionnaire, one-to-one correspondence was the most frequently accepted criterion by students who had taken a set theory course, whereas inclusion was most frequently chosen by those who had not taken such a course. When the questionnaire was readministered, more students of both kinds thought one-to-one correspondence was suitable for comparing the cardinality of infinite sets, but this was not accompanied by an increased tendency to reject other criteria. This supports Fischbein's claim that new scientific notions tend to coexist alongside primary intuitions, which are not easily replaced, as well as supporting his caution that teachers' verbal explanations alone do not suffice. Tsamir points to successful examples of cognitive-conflict teaching in the literature and suggests interventions based on sequences of comparison tasks, together with research to determine their impact on students' intuitions and performance. With this study, readers will be reminded that not only is teaching often not just telling, but also that incorrect intuitive (naive) notions about the way the world behaves can present mathematics teachers, just like physics teachers, with challenging pedagogical problems.

APOS Theory and Other Theoretical Frameworks

Although the next two papers are concerned with APOS theory, their aims are quite different. The first, by Kirk Weller, Julie Clark, Ed Dubinsky, Sergio Loch, Michael A. McDonald, and Robert R. Merkovsky, concentrates on gathering and analyzing research from a number of sources on students' performance and attitudes towards courses based upon APOS theory and the ACE teaching cycle. The second by David Meel is a reflective piece which considers recent views of mathematical understanding, but concentrates on comparing Pirie and Kieren's model for the growth of mathematical understanding with APOS theory.

Weller et al. collected and analyzed student performance and attitude findings from fourteen published and unpublished, mainly small-scale, studies on a variety of courses and topics taught using APOS theory and the ACE teaching cycle. These data come from courses on calculus and abstract algebra and from the teaching of the concepts of function, quantification, and induction, as well as from one longitudinal study on calculus and two affective studies. The article reports both comparative (with traditional students) and non-comparative studies of written and interview performance, as well as studies of the cognitive development of students in courses based on APOS theory.

The authors begin with an overview of APOS theory, followed by a description of their research framework that includes making an initial *genetic decomposition* of a concept, followed by the design of an instructional treatment using the ACE teaching cycle, and a subsequent analysis of the data collected during/after the treatment; this process may go through several iterations. For each concept, the ACE teaching cycle begins with activities (usually the writing of computer programs) to foster the development of those mental constructions suggested by the genetic decomposition, followed by class discussion to encourage reflection on those mental constructions, and ending with exercises to reinforce the ideas the students have constructed.

Next, the authors report on four prior calculus studies of Calculus, Concepts, Computers, and Cooperative Learning (C^4L) project and traditional students' understandings of the derivative, graphing, the chain rule, and sequences. In this section of their paper, as in subsequent sections, Weller et al. provide the actual interview tasks or descriptions of the indicators of understanding used, and report student performance data. For example, 71% of C^4L and 46% of traditional students were rated as demonstrating complete understanding of $f'(x)$ as the slope of the tangent to the graph at the point $(x, f(x))$. Turning to abstract algebra, three prior studies are described; results are given on how students performed on specific test and interview problems involving groups, subgroups, cosets, quotient groups, and permutation groups. For example, when asked, 71% of those in an APOS-based course provided "an acceptable definition of the method of representatives for cosets." The authors also report percentages of students' conceptions that were judged to be at the action, process, or object levels on various abstract algebra concepts.

Given the amount of detail presented by Weller et al., readers will be able to judge for themselves whether the performance and understanding of students in APOS-based courses is at a conceptually high level, as well as compare this with their own students' performance on similar tasks. When comparative findings (with traditional students) are provided for various tasks, the percentages of students in APOS-based courses that perform at a high level almost invariably exceeded the percentages of traditional students doing so.

The comparison statistics on students' attitudes indicates that students in APOS-based abstract algebra courses had more positive feelings toward the course than traditional students. Also, the authors discuss previously unreported results of an attitude questionnaire sent to students in eleven courses using various components of the APOS/ACE approach, as well as to an equal number of traditional students. Responses to this questionnaire favored the APOS/ACE group on all but two items.

Finally the authors note that many of the findings they present were gleaned from prior studies whose main focus was to uncover the nature of students' understandings, as elicited through written and interview responses to specially designed mathematics tasks. Still, they feel the findings paint a positive picture of the effectiveness of instruction based upon APOS theory and call for other researchers to conduct an independent, quantitative comparison study building upon these results.

In his contribution, David Meel offers mathematics education researchers and others an overview of various recent (post-1978) views of mathematical understanding. He first notes that subsequent to Richard Skemp's distinction between *relational understanding* and *instrumental understanding*, U. S. mathematics education researchers began to separate their ideas of knowledge from their ideas of understanding, with understanding focusing on the development of connections between concepts. However, there is still no single view of what constitutes mathematical understanding. Although Meel concentrates on a comparison of Pirie and Kieren's model and APOS theory, he briefly discusses several other theoretical frameworks. These include: cognitive, epistemological, and didactical obstacles (Bachelard, Cornu, Sierpinska), concept image and concept definition (Vinner), abstraction and generalization (Harel and Tall), multiple representations (Kaput), and operational and structural conceptions (Sfard).

Meel then provides a more detailed explication of Pirie and Kieren's eight layers for the growth of mathematical understanding beginning with *primitive knowing* and extending to *inventizing*. These eight layers are partitioned into four units by "don't need" boundaries that Meel sees as having parallels with the *action*, *process*, *object*, and *schema* building blocks of APOS theory, which is also described in detail. Next Meel uses Alan Schoenfeld's eight criteria for a theory, ranging from descriptive power, explanatory power, and scope to falsifiability, reliability, and multiple sources of evidence to do a point-by-point check that both Pirie and Kieren's model and APOS theory qualify as theories.

In drawing parallels between the two theories, Meel notes that both arise from a constructivist point of view with Pirie and Kieren's model having origins in Glasersfeld's perception of understanding as a constructive process of organizing one's knowledge structures and APOS theory being an extension of Piaget's ideas with a special emphasis on reflective abstraction. He also sees such elements as Pirie and Kieren's *folding back* as corresponding to Dubinsky's *de-encapsulation*. In addition, both theories can account for the development and surmounting of misconceptions and epistemological obstacles. Concept image has been explicitly linked to APOS theory's view of schema and Pirie and Kieren have expressed similar ideas. Meel goes on to draw parallels between the *Intra*, *Inter*, and *Trans* stages of schema development with Pirie and Kieren's layers of *image having*, *property noticing*, and *formalizing*.

In considering contrasting elements of the two theories, Meel notes they differ in the development of research questions. Those employing APOS theory first emphasize the mathematics, making and refining a genetic decomposition of a concept, like function or limit, to guide them in constructing their interview tasks, whereas Pirie and Kieren bring students' understandings to the foreground to direct their observations and interviews. Another difference is that Pirie and Kieren's model was developed while working with middle school and high school students, whereas APOS theory came from working with undergraduates.

Lastly, Meel comments on some implications of both theories for assessment and instruction. Although Pirie and Martin have developed a sequence for teaching linear equations using the model, mostly Pirie and Kieren have given recommendations to teachers, much like those in the National Council of Teachers of Mathematics *Principles and Standards for School Mathematics* (2000) regarding communication and problem solving, whereas Dubinsky and colleagues have developed the ACE teaching cycle and various undergraduate curricula. Pirie and Kieren emphasize the use of *provocative*, *invocative*, and *validating* questions to uncover and promote students' understandings, while researchers using APOS theory have proposed instructional sequences aimed at fostering certain mental constructions in students. In closing, Meel notes that his theoretical analysis raises some interesting questions. In particular, what similar, and what different, aspects would one uncover by examining the same interview session using first one, then the other, of these two theories?

Research Questions Regarding the Use of Technology

Jack Bookman and David Malone provide some preliminary results on student learning using technology when lessons are delivered via the Internet. They examined ten pairs of students using one of the Duke Connected Curriculum Project's study modules. These modules, for the first two years of college mathematics, include hypertext links, Java applets, sophisticated graphics, a computer algebra system, realistic scenarios, and questions requiring written answers. They are single-topic units meant to be completed in one to two hours and can be used as an integrated part of a course, for student projects, or as a supplement.

Pairs of students were videotaped and their computer output recorded. The authors then analyzed the tapes using grounded theory to develop categories, noting issues that seemed to facilitate or inhibit student learning. Bookman and Malone provide seven vignettes, illustrated with salient transcript excerpts. They observed students struggling with *Maple* syntax, being stuck and needing the instructor's intervention, ignoring hot links to relevant information, using paper and pencil in preference to *Maple*, confusing mathematical difficulties with *Maple* syntax problems, and trying to construct meaning out of unquestioned, yet incorrect, computer output. They also noted productive and unproductive discourse between the pairs.

Such observations led Bookman and Malone to raise questions for further research. Some of these are: What cognitive conditions prompt students to use hot links? How can one help students determine whether discrepancies are due to mathematical or technical errors? How, if a course is being delivered via the internet, could software developers simulate the role of the instructor? How can instructors help students recognize and differentiate between problems with the tool and problems of conceptual understanding? What are the implications of their observations for distance education? This preliminary study is part of the authors' long-term research agenda which aims to examine such questions in detail.

About *RCME*

A project of the AMS-MAA Joint Committee on Research in Undergraduate Mathematics Education (CRUME), the first volume of *RCME* appeared in 1994 under the editorship of Ed Dubinsky, Jim Kaput, and Alan Schoenfeld, who agreed to serve as editors for an unspecified length of time. Under their leadership four volumes of *RCME* were produced. This volume sees a transition to a new set of editors with Annie Selden, Guershon Harel, and Fernando Hitt being brought on board, and Ed Dubinsky agreeing to stay on for a bit. We would like to bid a fond farewell and extend thanks to Alan Schoenfeld and Jim Kaput for their previous service on the volumes. In addition, there has been a transfer of managerial responsibility for these volumes from CRUME to the MAA Special Interest Group on Research in Undergraduate Mathematics Education. We would like to acknowledge the financial and other support that this new affiliation brings to this and future *RCME* volumes.

We are also pleased to note the high regard with which *RCME* volumes are viewed in the wider mathematics education research community. Recently, the American Educational Research Association Special Interest Group on Research in Mathematics Education awarded Michelle Zandieh its Early Career Publication Award for outstanding mathematics education research for her *RCME IV* article.

And in conclusion, we would like to thank all those who have given of their time and expertise to review manuscripts for this and previous *RCME* volumes. Without these reviewers' help to the developing field of research in undergraduate mathematics education we could not produce the *RCME* volumes. And of course, we look forward to receiving many more manuscripts and to the production of future volumes.

Annie Selden
Ed Dubinsky
Guershon Harel
Fernando Hitt

CBMS Issues in Mathematics Education
Volume **12**, 2003

First-year Undergraduates' Difficulties in Working with Different Uses of Variable

María Trigueros and Sonia Ursini

ABSTRACT. Mathematics courses given at university level require a good understanding of algebraic concepts. In particular, it is essential to be able to work with different uses of variable. In this article we present the results of a study which investigated first-year undergraduates' capability to interpret and use variables as unknowns, general numbers and variables in simple functional relationships. The results are based on a detailed analysis of the responses given by 164 first-year college students to a questionnaire of 65 open-ended items and on data obtained when four students were interviewed. The results show the persistence of misconceptions and approaches characteristic of algebra beginners in secondary school. After several courses students do not seem to have a solid understanding of the concept of variable. Evidence is provided suggesting that students' responses are a reaction to key signs present in an expression (quadratic exponent, equal sign); and that students show a tendency to rely on and to apply memorized rules without examining their pertinence to a given context.

1. Introduction

The development of algebraic language and its use for different purposes requires the development of the concept of variable as a single multifaceted concept that includes different aspects. Researchers have emphasised the different uses of variable (e.g., Philipp, 1992; Schoenfeld & Arcavi, 1988). In particular, Usiskin (1988) stressed that different uses of variable are related to different conceptions of algebra, for example, generalized arithmetic, problem solving, study of relationships, study of structures. These different conceptions of algebra and the different uses of variable very often appear mixed together in the teaching of school algebra. Although students are usually taught how to use each, there is no emphasis on the particular aspects characterizing each use that might help students differentiate them. In particular, university students are expected to have acquired a solid and flexible handling of variable and the ability to distinguish between its different uses and to handle them in an integrated manner. Our concern is to investigate this assumption. Even though in the last twenty years much attention has been brought to secondary students' conceptions of variable (e.g., Bills, 2001; Booth,

We want to thank the editors and Dr. Kirk Weller for their valuable comments and suggestions that helped to improve this paper.

1984; Reggiani, 1994; Rosnik, 1981; Ursini, 1994; Wagner, 1981, 1983; Warren, 1999), considerably less effort has been dedicated to the study of the ways in which begining undergraduates work with this concept. Our study aims to contribute in this direction.

2. Brief historical antecedents

A starting point for investigating students' understanding of a mathematics concept consists, from our point of view, in making explicit what we mean by understanding that concept. This leads to the necessity of analyzing its mathematical characteristics and to relate them with what students are expected to do when working with situations involving it. Our first analysis of the concept of variable was based on a brief historical overview of the development of this concept and on our ownunderstanding of it as a multifaceted concept.

Analysis of some texts on the hi story of mathematics (Bell, 1945; Bunt, Jones & Bedient, 1976; Cajori, 1980; Gillings, 1982; Klein, 1968; Neugebauer, 1969; Rouse Ball, 1960; Youschkevitch, 1976) show that the presence of problems requiring the determination of a specific unknown quantity can be traced back to the earliest mathematical texts. The idea of change and one-to-one correspondence also appeared in different ancient cultures. For thousands of years mathematicians searched for useful methods for solving different problems related to these concepts. However, the general methods proposed were never expressed in a general way. Viète in the 16th century first expressed general methods symbolically by means of literal symbols representing general numbers. In the 17th century Fermat and Descartes, independently of each other, applied the new symbolism to geometry and presented the analytical method of expressing functions. Descartes considered the new algebra as a useful tool to model and think about problems implying undetermined or unknown quantities and also for expressing a dependence between variable quantities. This development was a cornerstone in the history of mathematics. Using these ideas as support, new mathematical and scientific theories have been developed. In all of them the concept of variable plays a fundamental role and at the same time this concept itself has been enriched with new and deeper meanings.

The introduction of symbols for variables is a culmination of a long historical process as well as a starting point of completely new disciplines. In the historical development of variable, its different uses were developed as different concepts that were later integrated and generalized so that it could be used in different areas of mathematics. At the present time the generic term "variable" subsumes the notions of specific unknown, general number, variables in functional relationship, and others. Therefore, the term "variable" as it is used today is a multifaceted concept, yet this fact is not acknowledged in many of the research projects related to the understanding of algebraic concepts nor in the teaching of algebra.

The conceptual richness of the notion of variable is lost when one analyses textbooks and the way in which this concept is introduced in school. A brief analysis of high school curricula and textbooks indicates that the more frequently used aspects of variable, when elementary algebra is taught, are the specific unknown, the general number, and variables in functional relationship. They are generally treated as facets of a single concept of variable, but without any explicit emphasis on these different uses.

TABLE 1. Categorization for the Concept of Variable

hline	Conceptualization and representation	Interpretation of symbols	Manipulation
Variable			
as specific unknown	of an unknown in a particular situation and/or in an equation.	as specific unknown in equations.	Factor, simplify, expand, transpose or balance an equation.
as general number	of a general number involved in general methods or rules deduced from numeric and/or geometric patterns and/or families of similar problems.	as a generalization in algebraic expressions or in the expression of general methods.	Factor, simplify and expand to rearrange expressions.
in functional relationships	of functional relationships (correspondence and variation) through tables, graphs or analytical representations.	representing correspondence and joint variation in analytical representations, tables and graphs.	Factor, simplify, expand to rearrange an expression, substitute values to determine intervals of variation, maximum/ minimum values and global behavior of the relationship.

3. A categorization for the concept of variable

Based on these antecedents and on our experience as teachers, we analyzed the content of some elementary algebra courses in order to highlight what we considered the basic capabilities needed for an understanding of the concept of variable. As a result of this analysis we developed a first theoretical framework that could help us in the study of students' conceptions of variable. This shown in Table 1.

Based on this framework an exploratory questionnaire consisting of 52 items (Quintero, Reyes, Trigueros, & Ursini, 1994; Quintero, Ursini, Reyes & Trigueros, 1995; Trigueros, Reyes, Ursini, & Quintero, 1996) was designed. In it we considered the three uses of variable mentioned above. For each of them, open ended questions requiring interpretation, manipulation and symbolization of variables were included. This questionnaire was given to 73 begining undergraduates, their responses were analyzed quantitatively and qualitatively (Trigueros, Ursini & Reyes 1996; Ursini & Trigueros, 1997). Based on this analysis the theoretical framework was refined to address in more detail specific aspects of the three uses of variable.

With respect to variable as specific unknown, we now consider that there are some basic aspects characterizing its understanding. First it is necessary to differentiate between situations or problems in which the value of the variable involved can be determined, and situations or problems involving unknown quantities in which the value or values cannot be determined. It is also necessary to be able to represent the unknown entities symbolically and to develop algebraic expressions that describe the relationships between given data. Eventually these expressions have to be linked in order to obtain an equation or a set of equations. When faced with equations, given or constructed, it is necessary to be able to manipulate the symbols involved in the equations in order to find the value or values of the specific unknown or unknowns that make the equation true. Moreover, it is also important to be able to substitute obtained values for the unknown in order to verify that they satisfy the given equation.

A prerequisite in order to develop an understanding of variable as general number, is the ability to recognize patterns and to find or deduce general rules and methods describing them. To be able to succeed in doing this it is necessary to distinguish the invariant aspects from the variable ones in problems related to a variety of situations that can involve geometrical or numerical sequences, or can be related to the structure of families of problems. It is necessary as well to be able to use symbols to represent general statements, rules or methods; to recognize a symbol as representing an indeterminate object; and to manipulate (expand, factor) expressions that involve general numbers without the need to assign them specific values. General numbers appear in expressions (e.g., $3x + 5$), tautologies (e.g., $2+x = x+2$), general formulae (e.g., $F = ma$) and as parameters in equations expressed in a generalized form (e.g., $ax + b = cx + d$). It is necessary to interpret them as general quantities and to distinguish them from the symbolic variables representing unknown quantities that can be determined.

In order to understand variables in functional relationships it is necessary to recognize situations where correspondence between variables and joint variation are present. These situations can involve information represented in tables, graphs, analytic expressions or verbal problems. For each one of these representations it is important to recognize correspondences between variables and to realize that a change in one variable may change another. The recognition of the relationship implies the capability to see that each variable can take many possible values depending on the interval where the relationship is defined. The ability to determine the correspondence between the variables is reflected by the capability to determine the value of one of the variables when the others are known. The ability to work with variation is reflected by the possibility to determine intervals or to find where a function is increasing or decreasing, is positive or negative, has a maximum or a minimum value, or other properties of the relationship. It is also necessary to be able to represent the information in terms of different representations and to be able to shift between them. When an analytical representation is required, it is necessary to symbolize related variables and to distinguish these expressions from equations and open expressions. It is also important to be able to manipulate the symbols to express the relationship in a convenient form, so that it is possible to deduce required properties.

A good understanding of variable implies the comprehension of all these aspects and the possibility to relate them with the corresponding use of variable. Students should develop the ability to interpret the different uses of variable as required by specific problem situations and to acquire the capability to shift between them in a flexible way, integrating them as facets of the same mathematical object. They should be able to manipulate symbols in order to perform simple calculations. These abilities would help them to develop a comprehension of why the operations work. The ability to symbolize rules and relationships in different problem situations would lead them to foresee the consequences of using variables. Beginning algebra students should gradually develop these basic capabilities that are related to a deep understanding of elementary algebra.

To summarize, for each of the uses of variable, we have stressed different aspects corresponding to different levels of abstraction at which it can be handled. These requirements can be presented in a schematic way as follows:

We consider that a student understands *variable as unknown* when he/she is able to:

> U1: recognize and identify in a problem situation the presence of something unknown that can be determined by considering the restrictions of the problem;
> U2: interpret the symbols that appear in an equation as representing specific values that can be determined by considering the given restrictions;
> U3: substitute for the variable the value or values that make the equation a true statement;
> U4: determine an unknown quantity that appears in equations or problems by performing algebraic and/or arithmetic operations;
> U5: symbolize the unknown quantities identified in a specific situation and use them to represent the situation by an equation.

It could be argued that an unknown is not a manifestation of a variable because it represents a fixed value; nevertheless, we consider that the first perception of the literal symbol when working in algebra is, or should be, that of a symbol representing any value, and that only in a second moment its role in the expression in which it appears can be defined. So, when presented with an expression (e.g., $4 + x^2 = x(x + 1) + 4$) we recognize that the variable represents a specific value only after having actually or mentally performed the necessary manipulations that allow us to recognize it as an equation and not, for example, as a tautology (e.g., $4 + x^2 = x(x + 1) - x + 4$).

We consider that a student understands *variable as a general number* when he/she is able to:

> G1: recognize patterns, perceive rules and methods in sequences and in families of problems;
> G2: interpret a symbol as representing a general, indeterminate entity that can assume any value;
> G3: deduce general rules and general methods by distinguishing the invariant aspects from the variable ones in sequences and families of problems;
> G4: manipulate (simplify, expand) the symbols;
> G5: symbolize general statements, rules or methods.

We consider that a student understands *variables in a functional relationship* (related variables) when he/she is able to:

> F1: recognize the correspondence between related variables independently of the representation used: tables, graphs, verbal expressions, analytic representations;
> F2: if possible, determine the values of a variable considered dependent given appropriate conditions on the variable considered independent;
> F3: if possible, determine the values of a variable considered independent given appropriate conditions on the variable considered dependent;
> F4: recognize the joint variation of the variables involved in a functional relationship;

F5: if possible, determine the range of one variable given the range of the other one;
F6: if possible symbolize a functional relationship based on the analysis of the data of a problem.

In some cases it is not possible to use F2, F3, F5 or F6 (consider, for example, the case of implicit functions). In these cases we think that a student should be able to use F1 and F4 together with the appropriate Us and Gs in order to obtain the information needed to solve the problem. This information is similar to that obtained when it is possible to use F2, F3, F5 or F6.

In contrast with previous research where the concept of variable is very often analyzed in terms of only one of its different uses or in terms of students' interpretation of literal symbols, this theoretical framework describes in detail the basic aspects involved in the understanding of variable as a multifaceted concept. It points to specific issues that can underlie students' difficulties with each of the uses of variable and so influence their success or failure in learning elementary algebra and more advanced mathematical topics.

This framework, or decomposition of variable, can be used as a tool to analyze students' strategies when they are solving algebra problems. It is useful as well for the design of questionnaires, interview protocols, observation grids and tests that can be of great value in the study of students' and teachers' difficulties and misunderstandings related to the concept of variable; for diagnosis; and for evaluation purposes. This decomposition can also serve as a guide in the design of teaching activities and sequences involving the concept of variable; in the analysis of problem situations where variables are involved, and in the analysis of textbooks or notes in order to find out how the concept of variable is introduced and developed through algebraic activities.

As part of our research project we analyzed begining undergraduates' understanding of the concept of variable in elementary algebra. In order to do this we designed a questionnaire together with interview and observation protocols based on the framework described above. This paper reports the results of the application of these instruments.

4. Study

The study consisted of a questionnaire and some interviews. The questionnaire was applied to the 164 first-year students enrolled in Economics, Administration, Accounting, Political Science and International Relations majors at a private Mexican university. All these degrees require at least six one semester courses of advanced mathematics (for example, calculus, linear algebra, probability, statistics). All the students that answered the questionnaire had completed at least four elementary algebra courses before entering the university. A common characteristics was that they all failed a classification test (a test that has been developed at this university and is used to select students that are considered able to follow a calculus course during the first semester), so they had not taken any advanced mathematics course. Additionally four students who obtained an average score when solving the questionnaire, were chosen to be interviewed.

Questionnaire. The questionnaire (see Appendix A for the English version and Appendix C for the Spanish version) includes 65 open-ended items. All of

TABLE 2. Questionnaire Analysis

	Specific unknown	General number	Variables in functional relationship
Number of questions	19	19	27
Mean	10.89	9.66	12.77
Standard deviation	2.95	3.9	4.55
Skew	0.355	−0.076	0.068
Reliability	0.73	0.77	0.79

them involved elementary mathematical notions. Each item emphasized one use of variable and was related with some specific aspects of its use as described in the theoretical framework (see Appendix B). There were sufficient items in the questionnaire so that all the aspects could be fully explored. Some of the questions were taken from questionnaires previously designed by other researchers (Küchemann, 1980; Mason, Graham, Pimm & Gowar, 1985; Ursini, 1994; Trigueros, Reyes, Ursini & Quintero, 1996).

Analysis of the questionnaire. An analysis in terms of Classical Test Theory (Muñiz, 1992) was used to verify the validity and internal consistency of the questionnaire as well as the difficulty and the power of discrimination of the questions.

The distribution of the scores was practically normal with a skew of 0.16. The coefficient of reliability obtained was 0.89, which indicates good internal consistency. The mean of the overall total scores on the questionnaire was 33.33 with a standard deviation of 9.79. This indicates that, on average, students answered about 50% of the questions correctly.

The difficulty of the questions, defined as the number of correct responses divided by the number of respondents, can be considered average because the most frequent indices oscillated between 0.4 and 0.6. The discrimination of the questions varied between −0.4 and 0.56, with a marked abundance between 0.43 and 0.55. Because this index can vary between −1 and 1, the discrimination of the items can be considered acceptable. A similar analysis was conducted taking the questions related to each one of the uses of variable as a sub-questionnaire. The analysis of the three sub-questionnaires is summarized in Table 2.

The results show that the reliability of each one of the three sub-questionnaires into which the questionnaire was divided is good and that the variance is greater in the parts concerned with the variable as generalized number and variables in functional relationships. The distribution for the variable as unknown is quite skewed. This seems to indicate that the students had less difficulty with this type of questions. This is also confirmed by the low standard deviation of this sub-questionnaire.

5. Analysis of students' performance

In order to have an overall idea of students' performance, highlighting their achievements and their failures, their responses to the questionnaire were analyzed from a quantitative and a qualitative perspective.

The percentage of correct answers was calculated for each question. Figure 1 shows the percentages of correct responses given to each question. This analysis

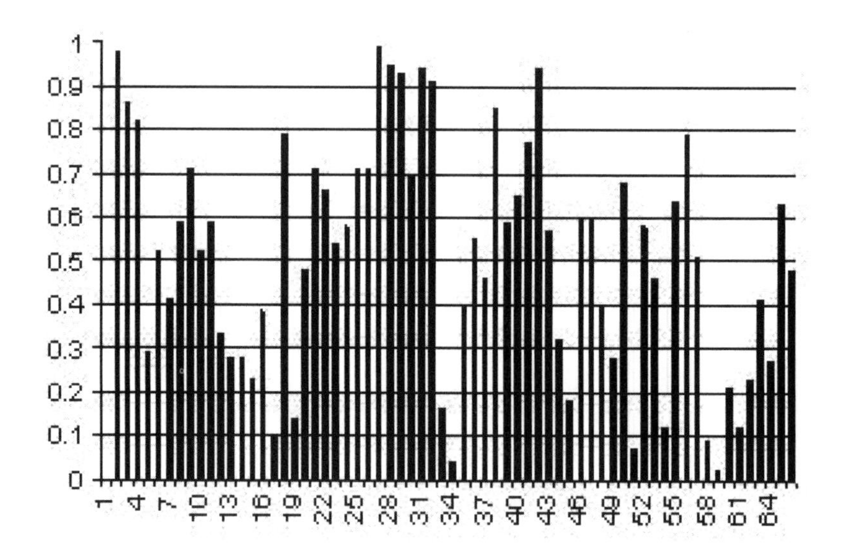

FIGURE 1. Proportion of correct responses

displays two features. No student answered all the questions correctly and no question was answered correctly by all students.

As can be seen in Figure 1, only seven questions (items 1, 26, 27, 28, 30, 31, 41) were answered correctly by more than 90% of the students. To correctly answer these items only very elementary algebraic notions are required. In particular, items 26, 27, 28, 30 and 31, were introduced to help students in the process of finding a specific pattern, and can be answered without any knowledge of algebra. Item 1 requires the translation of a simple sentence involving an unknown number into an algebraic expression, and item 41 requires the analysis of a table representing a linear relationship between speed and time, in order to focus students' attention on joint variation.

For questions related to each of the uses of variable, it was found that students had slightly less difficulty answering items involving variable as specific unknown. When calculating the ratio of the mean of correct answers to these questions and the total of questions involving this use, 0.57 was obtained, while the same ratio for questions related to variable as general number and functions were 0.51 and 0.47, respectively.

This overall analysis shows that these students' understanding of the different uses of variable is very weak. It also suggests that they will face a lot of difficulties when working with the more complex algebra problems that appear in advanced mathematics courses. What these students have learned in high school does not appear to allow them to work adequately with any of the different uses of variable.

Analysis with respect to the framework. In order to obtain a more complete idea of students' understanding of the notion of variable, their responses to the questionnaire were organized in three categories: no response, correct response, and incorrect response. We then analyzed all the incorrect responses obtained for

TABLE 3. Items Involving Variable as Specific Unknown

Item	No answer (%)	Correct (%)	Incorrect (%)	Examples of incorrect answers
6. How many values can the letter assume? $3 + a + a = a + 10$	16	41	43	3; several; 2; none; ∞
11. How many values can the letter assume? $7x^2 = 2x - 5$	24	33	43	1; several; none (without justifying their answer); ∞
12. How many values can the letter assume? $x/(x-4) = 3$	26	28	46	1; any value; none; R
14. How many values can the letter assume? $4+x^2 = x(x+1)$	31	23	46	2; ∞; several; R; N
16. Write the values that the letter can assume. $(x+3)^2 = 36$	17	10	73	$x = 3$
18. Write the values that the letter can assume. $10/(1+x) = 2$	18	14	68	$x = 2$
50. Write an expression to solve the following problem: The area of this figure (see Appendix I) is 27. Which is the area of the shaded square?	38	7	55	$a^2 + b^2 = 27; L = \sqrt{27}; 3(x) = 27; (L-3)(L-3); 27(\sqrt{27} \cdot 3 + \sqrt{27} \cdot 3); x - 3; 27 - 3^2$
52. Write an expression to solve the following problem: The cost for renting a car is \$25 per day plus \$0.12 per kilometer. How many kilometers can Diego drive in one day if he has \$40?	13	46	41	Arithmetical calculations followed by 125 as the result

each item together with the interviews. In this analysis we present the items which had more than 40% of incorrect answers, and we analyze these answers.

Tables 3 and 4 present the items involving variable as specific unknown and variable as general number respectively, to which more than 40% of students gave incorrect answers. The only exception is item 10. It involves variable as general number and was answered correctly by 59% of students. Item 10 was included in the table in order to compare the answers given to this item and the answers given to item 6 that involves variable as specific unknown. Both tables show the percentage of students giving no answer, correct and incorrect answers. Examples of the most frequent incorrect answers are presented in the rightmost columns of the tables. The symbols N and R refer to natural numbers and real numbers respectively.

The detailed analysis of responses that students had to the questionnaire and those given by the 4 students during the interviews, suggests difficulty in discriminating between variable as specific unknown and variable as general number. Evidence for this is provided by the answers given to items 6, 11, 12, 14 (see Table 3) and 5, 10, 13 (see Table 4). Observe that faced with general expressions (see Table 4) several students answer that the letter can assume only a finite number of values. In contrast, when faced with equations (see Table 3) they often answer that the

TABLE 4. Items Involving Variable as General Number

Item	No answer (%)	Correct (%)	Incorrect (%)	Examples of incorrect answers
4. Write a formula which means: An unknown number divided by 5 and the result added to 7.	3	29	68	$x/5 = y + 7$
5. How many values can the letter assume? $x + 2 = 2 + x$	4	52	44	1
10. How many values can the letter assume? $3 + a + a + a + 10$	15	59	26	1; 3
13. How many values can the letter assume? $(x + 1)^2 = x^2 + 2x + 1$	23	28	49	1; 2
22. To calculate the perimeter of a figure, sum the length of each one of its sides. Write the formula that expresses the perimeter of the following figure (see Appendix A).	5	54	41	$2(b+h)$; $(5+4)2$; $8+5\times 4$; $2(5)+2(4)$; $x+5+8+9 = P$; $x = 5L+L+x+x$; $8+5x\cdot 2$
32. How many points did you add to go from figure m to the figure following it? (See Appendix A.)	21	16	63	$m+n$; m^2-x^2; m ; $\#m + 2 + 2$; $2n$; ∞; $m + (m-1)$; $\#m +$ following#
33. Write a formula showing how points are added in order to obtain figure m. (See Appendix A.)	39	4	57	$m + m$; $y^2 - x^2$; $(figx)^2+2 = m$; $(1+(1+1))(2+(2+1))(3+(3+1))\dots(m+(m+1))$; $m+1 = (m+1)^2$; 1^2 , $2^2 \dots m^2$
34. Consider the following identities and complete the sequence: $1+2+3 = (3\times 4)/2$, $1+2+3+4 = (4\times 5)/2$, ... $1+2+3+ ... +n =$	11	40	49	$X\times n/2$; $(n\times n+1)/2$; $m\times n/2$; $5\times 6/2$; $13\times 14/2$; $(X+n)/(X-2)$

letter can assume infinitely many values. This suggests that students are having trouble with U2 and G2 (interpretation), which implies a feeble conceptualisation of both uses of variable. It is important to say that when evaluating students' answers to these questions we made the assumption that students were interpreting them as asking how many values of the variable make the equation true. This assumption is justified by the fact that it is always present in the teaching of equations at school algebra courses in our country.

Students' difficulties in interpreting the symbols clearly appear when the responses given to items 6 (Table 3) and 10 (Table 4) are compared. In item 6 the variable represents a specific unknown and in item 10 a general number. Students' answers reflected a consistent interpretation either of both expressions as open statements or as equations involving specific unknowns.

A superficial conceptualization of both variable as specific unknown and variable as general number is confirmed by the answers given to item 13 (Table 4). Here students seem to focus upon the fact that this expression represents an equation without seeing that the expressions on both sides of the equal sign are identical. As a result, they state that the situation will result in a specific value, or values, and they do not recognize that they have been given a tautology. Some of them answered that the variable involved could take only one single value, while others answered that the letter might assume two different values. This last answer could be induced by the presence of the quadratic exponent. This signals once more a failure with U2 and G2, which could result from an inability to manipulate (G4). For example, a typical response to item 13 during the interviews is illustrated by the following excerpt:

S: How many values? For this expression? (reads) $(x+1)^2 = x^2 + 2x + 1$, well, I don't remember how to do it, but there was a procedure . . . you have to use aprocedure, or a formula and you get two answers.
I: You mean the letter can assume two values? Why?
S: Yes, because it is quadratic.

Another item where we found that the quadratic exponent acted as a stimulus inducing students to treat the given equation as quadratic equation was item 14 (a linear equation with a "false" quadratic term, see Table 3). In this case students react to the presence of the exponent and do not realize that this equation is not quadratic but linear. This is also a manifestation of reluctance to manipulate (U4 and G4).

In general we can say that we found a tendency to give stereotyped answers based on definitions memorized when they were first introduced to the use of literal symbols at school. Usual responses during interviews were: "a letter stands for a number," "a letter represents any number," "it takes two values because it is a quadratic equation."

In order to differentiate between a variable representing a specific unknown and a general number it is necessary to manipulate expressions, actually or mentally, and obtain a form that might help recognize the use of variable involved. However, a strong tendency to avoid performing the necessary manipulations (U4 and G4) was found. Comments such as "3 is the solution" for equation $(x + 3)^2 = 36$ (item 16) were very common, and this kind of behavior was repeated while solving several of the items that asked for the values that the variable could take in a given expression, as for example in item 18.

This tendency to avoid manipulation was stronger when students were faced with quadratic equations (items 11, 12, 16, see Table 3). The great majority of students considered that the letter involved in these equations could assume only one value, as was illustrated for item 16. One student said it could assume any value, as illustrated in this excerpt from an interview about item 11:

I: How many values can the letter assume in $7x^2 = 2x - 5$?
S: Any value.
I: Why?
S: Because I am imagining that it is a parabola and the domain is all real numbers.

In this case, the student was considering the equation as the formula for a parabola, and not as an equation. She was reacting to the quadratic sign of the equation, but relating it to what she could remember from her course in Analytic Geometry. Later in the interview the student said:

S: When I have x squared, it is a parabola, that is what I remember I was taught in Analytic Geometry.

In most of these cases, students' answers were obtained by simple inspection. Once students found one appropriate value they considered the equation solved. All these answers confirm an avoidance of manipulation (U4).

The confusion between variable as general number and variable as unknown (U2 and G2) and the avoidance of manipulation (U4 and G4) were confirmed during the interviews when, in order to clarify our interpretation of their responses when faced with the equation $7x^2 = 2x - 5$ students were asked also how many values the letter can assume in $7x^2 + 2x - 5$. Their responses showed once more the difficulties in distinguishing between equations and open expressions:

I: How many values can the letter assume in $7x^2 + 2x - 5$?
S: Well, you have to find the numbers, you manipulate.
I: How?
S: I don't remember how, but you manipulate and there is a formula.

The difficulty in conceptualizing variables as specific unknowns (U1) was revealed by students' responses to items 50 and 52. In particular, the responses given to item 50 (see Table 1) highlight the difficulties students have in identifying and symbolizing the specific unknown in relation to the constraints of a problem. Although many of them tried to pose an equation symbolizing the specific unknown, the responses they give do not denote explicitly the specific unknown of the problem but another quantity involved in it. This is clearly a failure of U1. An example of this is shown in responses such as $(L-3)(L-3)$ where the student considers that the unknown is the side of the square obtained by completing the given picture with a small square of side 3 units. In this case, when the student was interviewed, it was clear that he was thinking about the side of the large square as the unknown instead of the side of the shadowed square. ". . .the total area is L^2, the area of the shaded part is $(L-3)(L-3)$," asked how he would get the value of L using his formula, he responds "that, I don't know."

Difficulty in conceptualizing variables as specific unknowns disappeared when students had to solve very simple problems (e.g., item 52, see Table 3). However, the responses given to item 52 highlight another aspect, namely, the persistence of arithmetic methods. Arithmetic methods are not bad in themselves, but algebra courses are supposed to introduce new ones, specifically those involving work with general statements. Although for item 52 students were explicitly asked to produce a formula for solving the problem, more than 40% gave a numeric answer without writing an equation. This is a partial failure of U5 (symbolization). In this case, the students guess a particular answer instead of considering the general aspects of the problem.

All these results suggest that a great number of university students are able to interpret, manipulate and symbolize variables as specific unknowns only at a very elementary level. They can recognize the presence of an unknown number in a problem, but their capabilities to process information are limited.

We have already mentioned some of the difficulties students have in interpreting a symbol as general number (G2), and in manipulating expressions (G4), but the analysis of students' responses shows that their understanding of this use of variable is very weak. Responses to item 4 show that even when students can symbolize and manipulate general numbers (they wrote, for example, $x/5$), they have problems when the expressions involved are a little bit more complicated. Many were not able to consider $x/5$ as a mathematical object that can be operated on to produce a new expression, for example, the expression $x/5 + 7$. The great majority needed to identify $x/5$ with a new general object, $x/5 = y$, and only then, they use it as an object in its own right. Moreover, after identifying $x/5$ with a new variable they did not consider necessary to separate the expressions $x/5 = y$ and $y + 7$ and wrote $x/5 = y + 7$. The combination of both expressions in a single one leads them to an incorrect result. The use of the equal sign to connect the solution steps can be considered as a reflection of their thinking process and as an evidence of their insecurity to assign a particular role to the variable in the problem. This is a clear example of failures of G4 and G5 (manipulation and symbolization):

During the interview a student explained why she wrote $x/5 = y+7$ as a response to this item:

S: A number divided by 5 . . . and you add 7 to the result. My number was $x/5$ equal to y, because I don't know the result . . . isn't it? Therefore the two literal symbols are different and I added 7 to the result.
I: Is it necessary to use the equal sign?
S: Yes, because y is the result of $x/5$ and it says here that we have to add that to 7.

The analysis of the responses given to item 22 (see Table 4) shows that students' capability to construct an expression which involves a general number slightly increased when the symbol was given and it had a clear reference (it represented the length of a segment) and the expression obtained represented the result of applying a known formula or algorithm. However, a tendency to use memorized formulae for the perimeter (for example, $2(b + h)$); to ignore the variable considering only the known longitudes (for example, $(5 + 4)2; 2(5) + 2(4)$); or to ignore the data and overgeneralize $(L + L + x + x)$, were also observed. These results suggest once more a rudimentary handling of variable as general number revealing a failure in interpreting a symbol as general number (G2) and in symbolizing (G5).

Finally, we want to stress that the capability to work with general numbers implies a process of generalization. Items 26 to 34 aimed to provide information about students' ability to generalize (interpret patterns, deduce general rules). In general, Mexican students have had some experience with exercises that ask them to continue a sequence of geometric or numeric patterns given only the first two or three steps. Most students in this sample were able to recognize the given pattern and follow a pattern when the first steps were explicitly given, that is, they were able to continue with the drawing and to write specific numbers in correspondence to the figures produced (items 26–28). Students could also find the points needed to go from one particular figure of the pattern to the next one (items 30, 31). However to answer items 32 and 33, for example, it is necessary to deduce a general rule from the observation of a sequence of figures representing square numbers, and express it analytically. The responses given to these items show that the great majority of students could not write an analytical expression representing a general rule. The following excerpt shows a typical conversation during the interviews, where even with help students were not able to deduce the rule they were asked for.

I: How many dots do you add when you go from figure m to the next one?
S: It is not possible to know.
I: Why?
S: Because I don't know the value of m.
. . .
I: Then from figure m to the next one . . .
S: It would be those of the next one minus those of figure m.
I: Try to write that down.
S: But I don't know how to write the next one.
I: Why don't you try something?
S: We cannot know.
I: Think about a way of expressing it.
S: All right, then n.
I: Why n?
S: Because n is next to m.

The response based on the order of the letters of the alphabet has already been found in studies with younger students (Booth, 1984). It is remarkable that we can find those same interpretations coming from college students. Several years of algebra courses have not succeeded in helping students in our study to overcome the most elementary difficulties.

The number of incorrect responses diminished for item 34, which requires finding the rule governing the addition of natural numbers. However, still a great number of students could not deduce a general rule and write the corresponding expression. These results show that students had difficulties with G1, G3 and G5 (recognizing patterns, deducing general rules, symbolizing).

The level of understanding of variable as general number is rudimentary as can be seen from this analysis. Although students were capable of recognizing a symbol as a representation of something indeterminate in straightforward cases, they appeared to have great difficulty when the activity required a slightly higher level of abstraction.

Responses to items involving variables in functional relationship that were answered incorrectly by more than 40% of students are presented in Tables 5 and 6. Items 38 and 49 are included for comparison purposes.

Regarding variables in a functional relationship we can say that students adequately handled the notion of correspondence between specific quantities (F1 and F2), independently of the representation used to express a relationship (table, graph, analytical representation). But they had difficulties with the idea of related variation as can be seen in their responses to items 57 and 58 (Table 6) which required the determination of intervals (F5). Analysis of students' responses showed that students tended to determine an interval by evaluating the function at the end points of the given interval, generalizing function's values within the interval without taking into account the specific characteristics of the function. This is the case, for example, for the response to item 57 given in an interview. The student had already found a quadratic relationship between the data given in the table:

TABLE 5. Items Involving Variables in Functional Relationships

Item	No answer (%)	Correct (%)	Incorrect (%)	Examples of incorrect answers
36. If $x + 3 = y$ which values can y have?	12	46	42	$x + 3$; N
38. Consider the table (see Appendix A) and write a general rule. Use n to represent the number of photocopies	11	65	24	n price; $6.25n = x$; $5n = 6.25$
43. Write a general rule that relates the numbers at the left side of the table to those at the right side (see Appendix A).	29	32	39	$10s = 30v$; $x + 1 = y + 1$; $1s = 3m/sec$; $x + m = y$ $d = v/t$
48. For each kilogram the tray shifts 4 centimeters. Find the relation between the weight of a product and the shift of the tray.	12	28	60	1 kg = 4 cm
49. For each kilogram the tray shifts 4 centimeters. If the tray shifts 10.5 cm, which is the weight of the product?	6	68	26	1 kg = 4 cm; 2.62; ? =10.5 cm
53. Observe the data (see Appendix A). Determine what happens with the value of y when the value of x increases?	10	11	79	increases; decreases

S: Well, the value of y would be between 256 and 1000, let me see, . . . (she writes $256 < y < 1000$, $256 < x^2 < 1000$) . . . then x must be between the root of 256 which is . . . can I use a calculator?

I: Yes, you can.

S: Aha, it is 16 and the root of 1000 is 100. That is all I need. So, x must be between 16 and 100.

I: Are you sure, even if it is a quadratic?

S: Yes, I am sure.

This kind of procedure is also illustrated by the responses given to item 58 (considering the interval [4, 676]), where an inappropriate generalization of the monotone nature of variation was made without taking into account the given relationship. The responses presented in Tables 5 and 6 show that students' difficulty in determining intervals substantially increased when they were required to determine the values of the independent variable (F3). Reversal thinking is needed when answering this type of problem, and this is known to pose difficulties to students (Ayers, Davis, Dubinsky, & Lewin, 1988). Responses to item 57 (Table 6) illustrate this difficulty. This item involves a quadratic relationship represented in a table. Observe that the percentage of incorrect answers together with the absence of response is over 90%, which is impressive if we consider that these students have already passed several algebra courses.

Students' difficulties with the notions of order and density of real numbers appeared again and again when they tried to determine the intervals in which

TABLE 6. Items Involving Variables in Functional Relationships

Item	No answer (%)	Correct (%)	Incorrect (%)	Examples of incorrect answers
57. Observe the data (see Appendix A). If we want the value of y to be between 256 and 10000, between which values must x be?	38	10	52	(16, 100)
58. Observe the data (see Appendix A). If x takes values between -2 and 26, between which values is y?	37	2	61	4 and 676
59. $40 - 15x - 3y = 17y - 5x$. Which value of y corresponds to $x = 16$?	34	21	45	$y = 10$; $y = -60$; $y = 6$; $y = 32$; $y = 13$; $y = 100$
62. For the given graph (see Appendix A), between which values of x to the values of y increase?	17	41	42	0 and 5 *and* 0 and -5; 0 and 5 *and* 0 and -2; $(0, 10)$; $(0, 5) \cup (0, \infty)$; -20 *to* $0 \cup 0$ *to* 5
63. For the given graph (see Appendix I), between which values of x, the values of y decrease?	16	27	57	-5 *and* -10; (4,∞); (∞ *and* -5)*and* (10, 20); 5 and 0 *and* -5 and -10; (5,∞); (4,∞); (∞ and -5)*and* (5,∞); (10 and 20)

a simple function increases or decreases. For example, to answer item 62 some students took a particular value of the independent variable as a reference and analysed the behavior of the dependent variable to its right and left, concluding that the values of the dependent variable increase both when the independent variable takes values between 0 and -5 and when it takes values between 0 and 5. The following excerpt illustrates this argument:

> I: Why do you say that y increases between 0 and -5 and from 0 to 5?
>
> S: I think that 0, x equals 0 and y equals 0 is the origin, where this . . . everything starts. Then when the curve moves upward for the values of y, I know that the values of y are increasing.

The symbolization of a functional relationship (F6) also presented serious difficulties to students (items 38, 43, 48 see Table 5). An example of this can be found in these two excerpts referring to item 38.

I: Can you write down a general rule that relates the number of photocopies with their cost?
S1: Well . . . this? (writes $xn = 6.25y$)
I: Why is that? Can you explain it to me?
S1: Well, what I think is that, at this side (indicating the first column of the table), if you multiply this by something, then you have to multiply the other side as well. And you multiply by 6.25 to know how much it will cost you . . .

S2: Then . . . I have . . . it is 1.25 (writes $n = 1.25$)
I: Is this the relationship? Can you explain it to me?
S2: Yes, because 6.25 divided by 5 is 1.25 and that is n.
I: Then, what is n?
S2: It is . . . it is the cost of one copy.
I: And is that the relationship between the number of copies and the cost?
S2: Yes, that is n.

The presence of a given symbol to represent a variable seemed to help students recognize and express symbolically a relationship between two variables. As can be seen in Table 5, a large number of students were able to symbolize a relationship when the symbol for the variable is given explicitly, as in item 38. This is not the case when the symbol is not given as, for example, in item 43. For this item only 32% of students could express the relationship analytically even though the relationship is very simple. This suggests that external support, such as a symbolized variable, is still needed and it can help focus students' attention on the main characteristics of the problem.

For these students arithmetic solution methods still prevailed over algebraic methods. This seemed to interfere with their capabilities to establish relationships and generalizations. When faced, for example, with word problems, such as item 48 (Table 5), a large number of students answered by writing 1kg = 4 cm. By means of this expression they tried to synthesise the information given in the problem using symbols. This can be viewed as a first, although unsuccessful, attempt to express symbolically the relationship between the given quantities. An analysis of the way students used the equals sign in this problem gives interesting information about their approach to it. For many of the students the equal sign assumes the role of a verb "displace" or "corresponds to." It was used as a tool supporting the analysis of the problem. This is shown in the following excerpts:

S1: One kilogram moves the tray 4 centimeters, then I have that 1 kg is equal to 4 cm.
S2: One kilogram is equal to 4 centimeters, then I use this and a rule of three.

This approach to the problem seemed to constrain many of the students to an arithmetic approach. This is confirmed by the fact that 68% of students correctly answered item 49 (Table 5). This demonstrates that their analysis helped them to recognize the relationship linking the given quantities and they could use it to perform the arithmetic calculations required to solve a particular instantiation of the problem. However, this kind of reasoning did not help them to generalize and symbolize the relationship. For item 48 (Table 5), only 28% of students were able to express the deduced rule analytically. In order to obtain a specific result the great majority used "the rule of three" (an algorithm to determine an unknown quantity linked by a proportional relationship to another quantity), a rule linked to arithmetic approaches for calculating the unknown of a problem.

Manipulation of algebraic expressions and relationships was the basis of many of the problems these students faced when working with related variables. The solution of item 59 (Table 6) requires the execution of various actions: substitution, transposition and grouping of similar terms. The responses given by students revealed a weakness in their ability to manipulate (G4) and a need to perform operations one by one following a specific order, writing down all the steps, without verifying if they are correct or not. The following is an excerpt from the interview. Faced with the expression $40 - 15x - 3y = 17y - 5x$ the student wrote:

$$\begin{aligned} 40 - 15x - 3y &= 17y - 5x \\ 40 - 20x &= 20y \\ 40 - 20(16) &= 20y \\ 40 - 320 &= 20y \\ y = \frac{-320}{20} &= 16 \end{aligned}$$

and concluded "it is sixteen."

In general the responses to items concerning relationships between variables show inconsistencies and inadequate generalizations even in the case of very simple problems. The majority of the students do not seem to conceive the relationship as a transformation process nor as a dynamic process of related variation (failure of F4). Their conception seemed to be limited to a static idea of correspondence between one value of the independent variable and one value of the dependent variable, where joint variation has no place.

6. Concluding remarks

The outcomes of this study provide evidence illustrating that, despite several years of study of algebra within the educational system, begining undergraduates' knowledge of the concept of variable is not significant. University students cannot differentiate between the different uses of variable. While they are capable of recognizing the role played by the variable in very simple expressions and problems, any small increase in complexity provokes inadequate generalizations. In particular strong evidence is given that students cannot distinguish between variable as a specific unknown and variable as general number and that they have serious difficulties when variables are related. Students' understanding of the concept of variable lacks the flexibility that is expected at this educational level.

When students had to work with more complex problems in which different uses of variable are involved, for example where they have to write and solve an equation or to deduce and work with a specific functional relationship, we found that they were not able to distinguish between them and to put them back together in order to be able to solve the original problem.

Summarizing, our results show that students' strategies are dominated by procedures which have not been interiorized, that is, they cannot use them without making explicit the chain of reasoning followed while solving the problem, and they need to rely on it throughout the whole process. In many of these procedures there is an evident confusion between the different uses of variable, particularly between variable as specific unknown and variable as a general number. Moreover, students were not capable of analyzing the steps they are following and to detect possible mistakes.

Despite the emphasis that most algebra courses give to manipulation, students did not seem to be able to perform it adequately. A strong tendency to avoid manipulation was found. When it was performed, the difficulties they still have clearly appeared. These could not be related to a specific use of variable.

Students' difficulties with joint variation when solving problems involving functional relationships were also evident. Although they were able to state that there is a relationship between quantities, they had plenty of difficulties in analyzing the joint variation of two related variables.

Difficulties in writing an equation were also detected. Although some of the problems were correctly understood and students were able to produce a correct result using arithmetic approaches, a great number of them were not able to use algebraic symbolism and shift to an algebraic way of facing the problem.

Superficial characteristics of the expressions seem to determine their decisions independently of the use of variable involved in the problem. Many of their actions appear to be provoked by external signs (for example, exponents, the equal sign, the way in which the question is posed) that lead them to respond in stereotyped forms. A similar phenomenon was observed in an entirely different context by Dubinsky and Yiparaki (2000). All the evidence we have suggests that begining university students are still anchored to an arithmetic way of thinking and to a very elementary conception of variable. Exposure to earlier algebra courses was not enough to help them shift from arithmetic to algebraic thinking and to develop a richer understanding of variable where all the aspects highlighted in the decomposition of this concept are integrated.

These findings are a call to re-think and analyze the way in which the concept of variable is taught in high school. If we want students to be prepared to work with mathematical concepts at the university level, where new and more complex uses of variable will be introduced and where solution methods rely on a solid understanding of elementary algebra, it is necessary to reconsider the way the concept of variable is taught in elementary algebra. Meanwhile it is necessary to design university courses that can foster students' understanding of the concept of variable. We hope this investigation invites other researchers to contribute with more research in this area to widen our comprehension of undergraduate students' understanding of algebra concepts.

The theoretical framework we designed to analyze students' difficulties with the concept of variable proved to be useful in the detailed analysis of students' strategies. As part of our research program we have used this framework to analyse textbooks (Ursini & Trigueros, 2001) and to study high school teachers' conceptions of variable (Juarez, 2002). Currently we are using it as a tool to design activities and materials to teach the concept of variable at different school levels.

References

Ayers, T., Davis, G., Dubinsky, E., & Lewin, P. (1988). Computer experiences in learning composition of functions. *Journal for Research in Mathematics Education, 19*(3), 243–259.

Bell, E. T. (1945). *The development of mathematics.* New York, London: McGraw-Hill.

Bills, L. (2001). Shifts in the meanings of literal symbols. In M. Van den Heuvel-Panhuizen (Ed.), *Proceedings of the 25th Conference of the International Group for the Psychology of Mathematics Education* (Vol. 2, pp. 161–168). Utrecht: Freudenthal Institute.

Booth, L. (1984). *Algebra: Children's strategies and errors.* Windsor: NFER-Nelson.

Bunt, L. N. H., Jones, P. S. & Bedient, J. D. (1976). *The historical roots of elementary mathematics.* Englewood Cliffs, NJ: Prentice-Hall.

Cajori, F. (1980). *A history of mathematics.* New York: Chelsea Publishing Company.

Dubinsky, E & Yiparaki, O. (2000). On student understanding of AE and EA quantification. In E. Dubinsky, A. H. Schoenfeld & J. Kaput (Eds.), *Research in collegiate mathematics education IV* (pp. 239–270). Providence, RI: American Mathematical Society.

Gillings, R.J. (1982). *Mathematics in the time of the pharaohs.* New York: Dover Publications.

Juarez, J. A. (2002). *La comprension del concepto de variable en profesores de secundaria.* MSc Thesis, Cinvestav, Mexico.

Klein, J. (1968). *Greek mathematical thought and the origin of algebra.* Cambridge: MIT Press.

Küchemann, D. (1980). *The understanding of generalized arithmetic (algebra) by secondary school children.* Unpublished doctoral dissertation, University of London.

Mason J., Graham A., Pimm D. & Gowar N. (1985). *Routes to/roots of algebra.* England: The Open University Press.

Muñiz, J. (1992). *Teoría clásica de los tests.* España: Ediciones Pirámide.

Neugebauer, O. (1969). *The exact sciences in antiquity.* New York: Dover Publications.

Philipp, R. (1992). The many uses of algebraic variables. *Mathematics Teacher, 85*(7), 557–561.

Quintero, R., Reyes, A., Trigueros, M., & Ursini, S. (1994). Misconceptions of variable. In D. Kirshner (Ed.), *Proceedings of the 16th Conference of the International Group for the Psychology of Mathematics Education, North American Chapter* (Vol. 1, p. 185). Baton Rouge, LA: Louisiana State University.

Quintero, R., Ursini, S., Reyes, A. & Trigueros, M. (1995). Students' approaches to different uses of variable. In L. Meira & D. Carraher (Eds.) *Proceedings of the 19th Conference of the International Group for the Psychology of Mathematics Education* (Vol. 1, p. 218). Brazil: Universidade Federal de Pernanbuco.

Reggiani, M. (1994). Generalization as a basis for algebraic thinking: Observations with 11–12 year old pupils. In J. P. Da Ponte & J. F. Matos (Eds.), *Proceedings of the 18th Conference of the International Group for the Psychology of Mathematics Education* (Vol. 4, pp. 97–104). Portugal: University of Lisbon.

Rosnik, P. (1981). Some misconceptions concerning the concept of variable. *Mathematics Teacher, 74*(6), 418–420.

Rouse Ball, W. W. (1960). *A short account of the history of mathematics.* New York: Dover Publications.

Schoenfeld, A. H. & Arcavi, A. (1988). On the meaning of variable. *Mathematics Teacher, 81*(6), 420–427.

Trigueros M., Reyes A., Ursini S. & Quintero R. (1996). Diseño de un cuestionario de diagnóstico acerca del manejo del concepto de variable en el álgebra. *Enseñanza de las Ciencias, 14*(3), 351–363.

Trigueros, M., Ursini, S. & Reyes, A. (1996). College students' conceptions of variable. In L. Puig & A. Gutierrez (Eds.), *Proceedings of the 20th Conference of the International Group for the Psychology of Mathematics Education* (Vol. 4, pp. 315-322). Spain: University of Valencia.

Ursini, S. (1994). *Pupils' approaches to different characterizations of variable in Logo.* Unpublished doctoral dissertation, University of London Institute of Education.

Ursini, S. & Trigueros M. (1997). Understanding of different uses of variable: A study with starting college students. In E. Pehkonen (Ed.), *Proceedings of the 21st Conference of the International Group for the Psychology of Mathematics Education* (Vol. 2, pp. 161-168). Finland: University of Helsinki.

Ursini, S. & Trigueros, M. (2001). A model for the uses of variable in elementary algebra. In M. Van den Heuvel-Panhuizen (Ed.), *Proceedings of the 25th Conference of the International Group for the Psychology of Mathematics Education* (Vol. 4, pp. 327-334). Utrecht: Freudenthal Institute.

Usiskin, Z. (1988). Conceptions of school algebra and uses of variables. In A. F. Coxford & A. P. Shulte (Eds.), *The ideas of algebra K–12* (pp. 8–19). Reston, VA: National Council of Teachers of Mathematics.

Wagner, S. (1981). An analytical framework for mathematical variables. In *Proceedings of the 5th Conference of the International Group for the Psychology of Mathematics Education* (pp. 165-170). Grenoble, France.

Wagner, S. (1983). What are these things called variables? *Mathematics Teacher 76*(7), 474–479.

Warren, E. (1999). The concept of variable: gauging students' understanding. In O. Zaslavsky (Ed.), *Proceedings of the 23rd Conference of the International Group for the Psychology of Mathematics Education* (Vol. 4, pp. 313-320). Haifa: Technion, Israel Institute of Technology.

Youschkevitch, A. P. (1976). The concept of function up to the middle of the 19th century. *Archives of History of Exact Sciences, 16*(1), 37–85.

Appendix A

Questionnaire

In the following exercises do not calculate the number, write only a formula.

1. Write a formula which means: An unknown number multiplied by 13 is equal to 127.
2. Write a formula which means: An unknown number multiplied by the sum of the same unknown number plus 2 is equal to 6.
3. Write a formula which means: An unknown number is equal to 6 plus another unknown number.
4. Write a formula using symbols which means: An unknown number divided by 5 and the result added to 7.

How many values can the letter assume in the following expressions?

5. $x + 2 = 2 + x$
6. $3 + a + a = a + 10$
7. $x = x$
8. $4 + s$
9. $x + 5 = x + x$
10. $3 + a + a + a + 10$
11. $7x^2 = 2x - 5$
12. $\frac{x}{x^2-4} = 3$
13. $(x+1)^2 = x^2 + 2x + 1$
14. $4 + x^2 = x(x+1)$

Write the values that the letter can assume:

15. $13x + 27 - 2x = 30 + 5x$
16. $(x+3)^2 = 36$
17. $4 + x = 2$
18. $\frac{10}{1+x^2} = 2$

Reduce the following expressions to equivalent ones:

19. $(x^2+1)(x^2-2) =$
20. $a + 5a - 3a =$
21. $y^2 + 2y + 4y^2 - 5y - 8 =$
22. **To calculate the perimeter of a figure, sum the length of each one of its sides. Write a formula that expresses the perimeter of the following figure:**

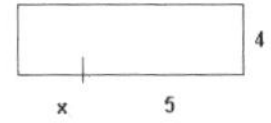

Write formulas to calculate the area of the following figures:

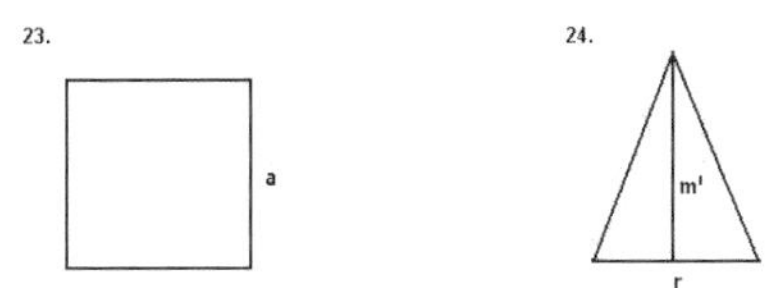

25. **In the following figure, the polygon is not entirely visible. Since, we do not know how many sides the polygon has, let say it has N sides. The length of each side is 2 centimeters. Write a formula to calculate the perimeter of the polygon.**

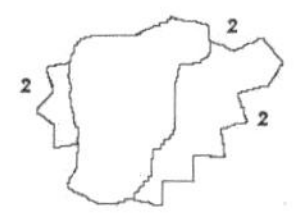

Consider the following figures:

Figure 1	Figure 2	Figure 3	Figure 4
number of dots: 1	number of dots: 4	number of dots: 9	number of dots:

26. How many dots does Figure 4 have?
27. Draw Figure 5 and give the number of dots.
28. Draw Figure 6 and give the number of dots.
29. Assume you can keep on drawing figures until Figure m. How many dots does Figure m have?

To complete the figures above you were adding dots,

30. How many dots did you add to go from Figure 1 to Figure 2?
31. How many dots did you add to go from Figure 2 to Figure 3?
32. How many dots did you add to go from Figure m to the figure following it?
33. Write a formula showing how dots are added in order to obtain Figure m.
34. **Consider the following identities and complete the sequence:**

$1+2+3=\frac{(3\times 4)}{2}$
$1+2+3+4=\frac{(4\times 5)}{2}$
$1+2+3+4+\ldots+n=$

If $x+3=y$

35. Which values can x have?
36. Which values can y have?

If $y=7+x$

37. What happens to the values of y when the value of x increases?

In order to facilitate his work, an employee has made the following table.

Number of photocopies	*Cost*
5	6.25
10	12.30
15	
	25.00
25	31.25
35	
	62.50
100	

38. Complete the table above.
39. Consider the table above and write a general rule relating the number of photocopies and the cost. Use n to represent the number of photocopies.

Consider the following table.

Seconds	*Speed*
0	0 m/s
10	30 m/s
15	
20	60 m/s
35	
50	
60	

40. Complete the table above.

Consider the table above and answer:

41. If the time increases, does speed increase or decrease?

42. On a sheet of paper draw a coordinate system, mark the dots at which coordinates in the table appear, and join them by drawing a curve.
43. Write a general rule that relates the numbers at the left side of the table to those at the right side.

Consider $y = 3 + x$

44. If we want the values of y to be bigger than 3 but smaller than 10, which values can x take on?
45. If the values of x are between 8 and 15, which are the values corresponding to y?

Consider the following expressions: $n + 2$ and $2xn$

46. Which is bigger?
47. Explain your answer.

In a market, a platform scale is used. For each kilogram the tray shifts 4 centimeters.

48. Find the relationship between the weight of a product and the shift of the tray.
49. If the tray shifts 10.5 cm, which is the weight of the product?

Write formulas to solve the following problems:

50. The area of this figure is 27. Which is the area of the shaded square?

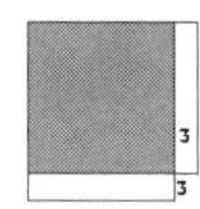

51. Juan is 15 years older than Santiago. The sum of their ages is 41. What are Juan's and Santiago's ages?
52. The cost for renting a car is \$25 per day plus \$0.12 per kilometer. How many kilometers can Diego drive in one day if he has \$40?

Consider the table and answer the questions:

x	y
0	0
10	100
−15	225
25	625
20	400
−10	100
15	225
−20	400

53. Determine what happens with the value of y when the value of x increases.
54. For which value of x, does y achieve its maximum value?
55. For which value of x, does y achieve its minimum value?
56. Write a general rule that relates variable x to variable y.
57. If we want the value of y to be between 256 and 10000, between which values x must be?
58. If x takes values between −2 and 26, between which values is y?

Given the expression $40 - 15x - 3y = 17y - 5x$

59. Which value of y corresponds to $x = 16$?
60. Between which values must be x, if we want the value of y to be between 1 and 5?
61. Assume that x takes values between −5 and 5, for which value of x does y reach its maximum value?

Given the graph:

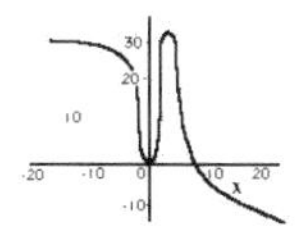

62. Between which values of x do the values of y increase?
63. Between which values of x do the values of y decrease?

64. For which value of x does y reach its maximum value?
65. For which value of x does y reach its minimum value?

Appendix B

Items' relationship with theoretical framework by item number

1. Symbolization of *variable as unknown*. Producing a simple expression requiring a product operation. U5

2. Symbolization of *variable as unknown*. Producing an expression in which the variable appears several times. The distributive law is required. U5

3. Symbolization of *variable in a functional relationship*. Producing an expression in which the variable appears in both sides of the equal sign. The expression contains an addition operation. F6

4. Symbolization of *variable as general number*. Producing an expression in which the variable appears in a fraction. An addition operation is required. G5

5. Interpretation of *variable as general number* in a tautological expression. G2

6. Interpretation of *variable as unknown* in a linear equation in which the variable appears several times and an addition operation is involved. U2

7. Interpretation of *variable as general number* in a tautology. G2

8. Interpretation of *variable as general number* in an expression in which an addition operation is involved. G2

9. Interpretation of *variable as unknown*, in a linear equation in which the variable appears on both sides of the equal sign. U2

10. Interpretation of *variable as general number* in an expression in which the variable appears several times. G2

11. Interpretation of *variable as unknown* in a quadratic equation. U2

12. Interpretation of *variable as unknown* in a quadratic equation. The quadratic term appears in the numerator. U2

13. Interpretation of *variable as general number* in a tautological expression involving quadratic terms. G2

14. Interpretation of *variable as unknown* in a linear equation involving a quadratic term that disappears after manipulating the expression. U2

15. Determination of the *unknown* value that appears in a linear equation in which the variable appears several times. Manipulation is required. U2, U3, U4

16. Determination of the *unknown* value that appears in a quadratic equation. Manipulation is required. U2, U3, U4

17. Determination of the *unknown* that appears in a linear equation. The solution is a negative number. U2, U3, U4

18. Determination of the *unknown* value that appears in a quadratic equation. The quadratic term appears in the numerator. Manipulation is required. U2, U3, U4

19. Manipulation of *variable as general number* in an expression in which the variable appears squared several times. Addition and multiplication of the variable are involved. G4

20. Manipulation of *variable as general number* in an expression in which the variable appears several times. Addition is involved. G4

21. Manipulation of *variable as general number* in an expression in which the variable appears several times and sometimes squared. Addition is involved. G4

22. Symbolization of *variable as general number*. Producing an expression considering the data given in relation to a geometric figure. G2, G5

23. Symbolization of *variable as general number*. Producing an expression considering the data given in relation to a geometric figure. G2, G5

24. Symbolization of *variable as general number*. Producing an expression considering the data given in relation to a geometric figure. G2, G5

25. Symbolization of *variable as general number*. Producing an expression considering the data given in relation to a geometric figure. Generalization is required. G2, G3, G5

26. Recognizing a pattern. Making explicit the first steps of a pattern. G1

27. Recognizing a pattern. Making explicit the first steps of a pattern. G1

28. Recognizing a pattern. Making explicit the first steps of a pattern. G1

29. Symbolization of *variable as general number* as a result of recognizing a pattern with a given symbol. G3, G5

30. Recognizing a pattern. Making explicit the first steps to find a pattern. G1

31. Recognizing a pattern. Making explicit the first steps to find a pattern. G1

32. Symbolization of *variable as general number* as a result of recognizing a pattern, with a given symbol. G3, G5

33. Symbolization of *variable as general number* as a result of recognizing a pattern with a given symbol. G3, G5

34. Symbolization of *variable as general number* as a result of recognizing a pattern with a given symbol. G3, G5

35. Interpretation of *variable in a functional relationship* focusing on the domain. F1

36. Interpretation of *variable in a functional relationship* focusing on the range. F1

37. Interpretation of *variable in a functional relationship* focusing on joint variation of variables and on the range. F1, F4

38. Completing a table focusing on correspondence. F1, F2, F3

39. Symbolization of a *functional relationship* using a given symbol and based on data given in a table. F6

40. Completing a table focusing on correspondence. F1, F2

41. Interpretation focusing on joint variation and on the domain of *variables in functional relationship* represented by a table. F4

42. Drawing a graph to represent a *functional relationship* using data given in a table. F1

43. Symbolization of a *functional relationship* using a given symbol and based on data given in table. F6

44. *Variables in functional relationship.* Determining an interval for the dependent variable. The relationship involved is a linear one. F1, F5

45. *Variables in functional relationship.* Determining an interval for the independent variable. The relationship involved is a linear one. F1, F5

46. Interpretation of *variable as general number*, comparing two expressions. G1, G2

47. Interpretation of *variable as general number*, comparing two expressions. G1, G2

48. Symbolization of a *functional relationship* from a verbal problem, when the symbol is not given. F1, F6

49. *Variables in functional relationship.* Determining the value of the dependent variable given the value of the independent one, from an analytical expression. F1, F2

50. Symbolization of a *variable as unknown* in order to write an equation, given a verbal problem. U1, U5

51. Symbolization of a *variable as unknown* in order to write an equation, given a verbal problem. U1, U5

52. Symbolization of a *variable as unknown* in order to write an equation, given a verbal problem. U1, U5

53. Interpretation of *variable in a functional relationship* from pairs of numbers presented in a table not ordered by first coordinate, focusing on joint variation. The data represent a quadratic relationship. F4

54. Interpretation of *variable in a functional relationship* from pairs of numbers presented in a table not ordered by first coordinate. The data represent a quadratic relationship. The item requires finding a specific value of the dependent variable and the corresponding value of the independent one. F2

55. Interpretation of *variable in a functional relationship* from pairs of numbers presented in a table, not ordered by first coordinate. The data represent a quadratic relationship. The item requires finding the specific value of the dependent variable and the corresponding value of the independent one. F2

56. Symbolization of a *functional relationship* from pairs of numbers presented in a table not ordered by first coordinate. The expression represents a quadratic relationship, the symbols are given. F6

57. Interpretation of *variable in a functional relationship*, from pairs of numbers presented in a table not ordered by first coordinate. The data represent a quadratic relationship. The item requires determining an interval for the independent variable given an interval for the dependent one. F3, F5

58. Interpretation of *variable in a functional relationship*, from pairs of numbers presented in a table, not ordered by first coordinate. The data represents a quadratic relationship. The item requires determining an interval for the dependent variable given an interval for the independent one. F2, F5

59. Manipulation of *variables in functional relationship* when an analytical representation with more than one appearance of the dependent and the independent variables is given. The item requires substituting a given value for the independent variable and calculating the corresponding value for the dependent one. G4, F2

60. Interpretation of *variable in a functional relationship* when an analytical representation with more than one appearance of the dependent and the independent variables is given. The item requires determining an interval for the independent variable, given an interval for the dependent one. F3, F5

61. Interpretation of *variable in a functional relationship* when an analytical representation with more than one appearance of the dependent and the independent variables is given. The item requires determining the maximum value of the dependent variable within a given interval for the independent variable. F2

62. Interpretation of *variable in a functional relationship* from a graph. The focus is on joint variation in intervals. F5

63. Interpretation of *variable in a functional relationship* from a graph. The focus is on joint variation in intervals. F5

64. Interpretation of *variable in a functional relationship* from a graph. The item requires determining the specific value of the independent variable for which the dependent variable reaches its maximum. F3

65. Interpretation of *variable in a functional relationship* from a graph. The item requires determining the specific value of the independent variable for which the dependent variable reaches its minimum. F3

Appendix C

Cuestionario

En este ejercicio, solamente escribe una fórmula. NO CALCULES **el número.**

1. Escribe una fórmula que exprese: Un número desconocido multiplicado por 13 es igual a 127.
2. Escribe una fórmula que exprese: Un número desconocido multiplicado por la suma del mismo número desconocido con 2 es igual a 6.
3. Escribe una fórmula que exprese: Un número desconocido es igual a 6 más otro número desconocido.
4. Escribe una fórmula que exprese: Un número desconocido dividido por 5 y el resultado sumado a 7.

Para cada una de las siguientes expresiones ¿cuántos valores puede tomar la letra?

5. $x + 2 = 2 + x$
6. $3 + a + a = a + 10$
7. $x = x$
8. $4 + s$
9. $x + 5 = x + x$
10. $3 + a + a + a + 10$
11. $7x^2 = 2x - 5$
12. $\frac{x}{x^2-4} = 3$
13. $(x + 1)^2 = x^2 + 2x + 1$
14. $4 + x^2 = x(x + 1)$

Para cada una de las siguientes expresiones escribe los valores que puede tomar la letra:

15. $13x + 27 - 2x = 30 + 5x$
16. $(x + 3)^2 = 36$
17. $4 + x = 2$
18. $\frac{10}{1+x^2}$

Reduce las siguientes expresiones a una equivalente:

19.$(x^2 + 1)(x^2 - 2) =$

20. $a + 5a - 3a =$
21. $y^2 + 2y + 4y^2 - 5y - 8 =$
22. **El perímetro de una figura se calcula sumando la longitud de sus lados. Escribe la fórmula que expresa el perímetro de la siguiente figura:**

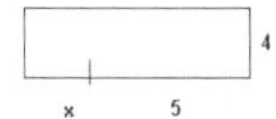

Escribe una fórmula para calcular el área de las siguientes figuras:

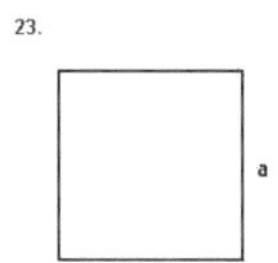

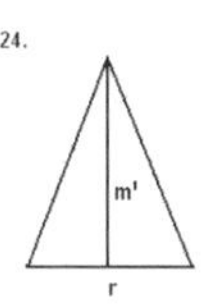

25. **En la siguiente figura, el polígono no es completamente visible. Debido a que no sabemos cuántos lados tiene el polígono en total diremos que tiene N lados. Cada lado mide 2 centímetros de longitud. Escribe una fórmula para calcular el perímetro del polígono.**

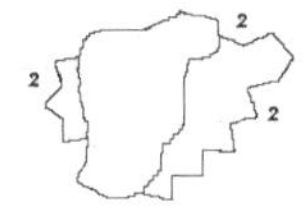

Observa las siguientes figuras:

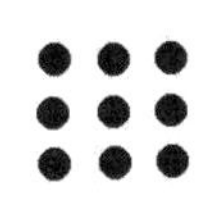

Figura 1	Figura 2	Figura 3	Figura 4
número de puntos: 1	número de puntos: 4	número de puntos: 9	número de puntos:

26. ¿Cuántos puntos hay en la Figura 4?
27. Dibuja la Figura 5 y da el número total de puntos.
28. Dibuja la Figura 6 y da el número total de puntos.
29. Imagínate que puedes seguir dibujando figuras hasta la Figura m. ¿Cuántos puntos en total tendrá la Figura m?

Si para hacer las figuras del ejercicio anterior vas agregando puntos

30. ¿Cuántos puntos agregas para pasar de la Figura 1 a la Figura 2?
31. ¿Cuántos puntos agregas para pasar de la Figura 2 a la Figura 3?
32. ¿Cuántos puntos agregas para pasar de la figura #m a la siguiente?
33. Escribe una fórmula que muestre como vas agregando puntos hasta llegar a la Figura m
34. **Observa las siguientes identidades y completa la secuencia:**

$1 + 2 + 3 = \frac{(3\times 4)}{2}$

$1 + 2 + 3 + 4 = \frac{(4\times 5)}{2}$

$1 + 2 + 3 + 4 + \ldots + n =$

Si $x + 3 = y$

35. ¿Qué valores puede tomar x?
36. ¿Qué valores puede tomar y?

Si $y = 7 + x$

37. ¿Qué les pasa a los valores de y cuando los valores de x aumentan?

Para evitarse estar haciendo multiplicaciones el empleado de una papelería ha elaborado la siguiente tabla.

Númer o de copias	*precio*
5	6.25
10	12.30
15	
	25.00
25	31.25
35	
	62.50
100	

38. Complétala.
39. Considera la tabla anterior y escribe una regla general que relacione el número de copias con el precio. Usa n para representar el número de copias.

Observa la siguiente tabla.

segund os	*velocidad*
0	0 m/s
10	30 m/s
15	
20	60 m/s
35	
50	
60	

40. Complétala.

Considera la tabla anterior y contesta

41. Si aumenta el tiempo ¿la velocidad aumenta o disminuye?
42. En una hoja aparte dibuja un sistema de coordenadas, marca los puntos cuyas coordenadas aparecen en la tabla y únelos trazando aproximadamente una curva.
43. Escribe una regla general que asocie a los números de la izquierda los números de la derecha.

Considera la expresión $y = 3 + x$.

44. Si queremos que los valores de y sean mayores que 3 pero más pequeños que 10 ¿qué valores puede tomar x?
45. Si x toma valores entre 8 y 15 ¿qué valores corresponden a y?

De las siguientes dos expresiones $n + 2$ y $2xn$

46. ¿Cuál es más grande?
47. Justifica tu respuesta.

El peso de la mercancía que se compra en el mercado se mide con una báscula. Por cada kilogramo de peso la charola de la báscula se desplaza 4 cm.

48. Encuentra la relación entre el peso de la compra y el desplazamiento de la charola.
49. Si la charola se desplaza 10.5 cm ¿cuántos kilos pesa la mercancía?

Escribe una fórmula para resolver los siguientes problemas:

50. El área total de la figura es 27. Calcula el lado del cuadrado sombreado.

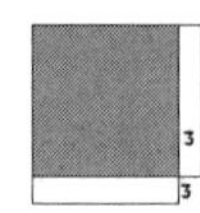

51. Juan es 15 años mayor que Santiago. La suma de las dos edades es 41. ¿Cuáles son las edades de Juan y Santiago?
52. Rentar un automóvil cuesta \$25por día, más \$0.12 por kilómetro. ¿Cuántos kilómetros puede manejar Diego en un día, si sólo dispone de \$40?

Observa los datos siguientes

x	y
0	0
10	100
-15	225
25	625
20	400
-10	100
15	225
-20	400

53. Determina qué pasa con el valor de y cuando el valor de x va creciendo
54. ¿Para qué valor de x, alcanza y su valor máximo?
55. ¿Para qué valor de x, alcanza y su valor mínimo?
56. Escribe una regla general que relaciona a la variable x con la variable y.
57. Si queremos que el valor de y esté entre 256 y 10000 ¿entre qué valores tiene que estar x?
58. Si x toma valores entre -2 y 26 ¿entre qué valores estará y?

Dada la expresión $40 - 15x - 3y = 17y - 5x$

59. ¿Cuál es el valor de y para $x = 16$?
60. Para que el valor de y esté entre 1 y 5 ¿entre qué valores debe estar x?
61. Supón que x toma valores entre -5 y 5, ¿para qué valor de $x$alcanza y su valor máximo?

Dada la siguiente gráfica:

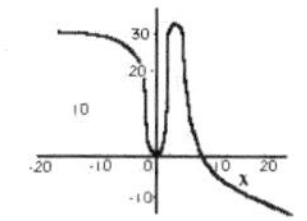

medskip

62. ¿Entre qué valores de x, los valores de y crecen?
63. ¿Entre qué valores de x, los valores de y decrecen?
64. ¿Para qué valor de x, se obtiene el valor máximo de y?
65. ¿Para qué valor de x se obtiene el valor mínimo de y 62. ¿Entre qué valores de x, los valores de y crecen?
63. ¿Entre qué valores de x, los valores de y decrecen?
64. ¿Para qué valor de x, se obtiene el valor máximo de y?
65. ¿Para qué valor de x se obtiene el valor mínimo de y

Dapartamento de Matemáticas, ITAM, México
E-mail address: `trigue@itam.mx`

Departament de Matemática Educativa, CINVESTAV, México
E-mail address: `sursinil@conacyt.mx`

CBMS Issues in Mathematics Education
Volume **12**, 2003

Cooperative Learning in Calculus Reform: What Have We Learned?

Abbe Herzig and David T. Kung

ABSTRACT. Research on calculus reform has shown some promising results about the achievement, attitudes, and persistence of students in reformed calculus classes. However, "reform calculus" embodies a broad range of classroom innovations, including the use of technology and significant changes to curriculum and instruction. Many of the reform projects reported in the literature combine a variety of these elements, and virtually no research has attempted to differentiate among their effects. In addition, most studies exhibit a number of problems that limit the conclusions that can be drawn from them. As a result, departments seeking to institute reforms have limited evidence to guide their decisions.

In this paper, our focus is on research that assesses the effects of cooperative learning on student outcomes in reformed calculus classrooms. We focus on cooperative learning as a more specific, controlled element of calculus reform than many of the broad-scale research projects reported in the literature. We describe a study completed at the University of Wisconsin at Madison that was designed to examine the effects of a specific type of cooperative learning on student performance, attitudes, and persistence in a first-semester calculus course. Aside from the specific results discussed, this study is also presented as an attempt at controlled research in which we try to address some of the limitations of previous research. We conclude with suggestions for future research about calculus reform.

Introduction

The calculus reform movement has exploded onto the college mathematics scene over recent years, revolutionizing the way the subject is taught at many institutions, although not without controversy. Most reform efforts involve complex sets of classroom changes, including syllabus changes, the use of calculators and computers, multiple representations of mathematical concepts, student projects and communicating about mathematics, and group work (Ross, 1996). However, research on these reforms has lagged behind the reforms themselves. Virtually no

The authors would like to thank Robert Wilson, Joel Robbin, and the Mathematics Department at the University of Wisconsin for their generous support, and the teaching assistants and students who participated in the study. Thanks also to Margaret Meyer whose thoughtful advice helped improve the quality of this paper, and to the reviewers and editors for their helpful comments.

research has attempted to differentiate which changes have which effects, with the result that departments seeking to institute reforms have limited evidence to guide their decisions.

Because of the diversity and quantity of calculus reform efforts, a complete review of the literature is beyond the scope of this paper (for a general overview, see Darken, Kuhn, & Wynegar, 2000; Tucker & Leitzel, 1995; Tucker, 1990). Instead, we will try to untangle what the research can tell us about one prominent feature of calculus reform: cooperative learning. Cooperative learning is a complex construct in and of itself, encompassing a broad range of pedagogies, and requiring significant investments in terms of instructor training, student training, development of curriculum and assessment materials. This focus will allow us a more detailed look at the research. In addition, many of the issues surrounding research on cooperative learning apply more generally to other areas of reform. We begin with an overview of the history of cooperative learning in mathematics and the calculus reform movement, and then provide a review and critique of the relevant research literature. We then present an example of a research project that attempts to address some of the gaps in the existing research.

Background

History. Cooperative learning had been part of American education since the early nineteenth century, beginning with the common school movement of the early 1800s and the progressive schools advocated by John Dewey in the earlier part of this century. This changed in the depression of the early 1930's, when business interests organized to advocate interpersonal competition in the schools (Johnson & Johnson, 1994a). Competition among students soon became entrenched in the American educational system, so that by the early 1960's competition came to be viewed as the standard way of structuring student interactions in the classroom (Johnson, Johnson, & Smith, 1991).

In the late 1950's, when the Soviet Union launched Sputnik and the space race began, pressure was placed on the American educational system to train more students in mathematics and science. As the baby boom generation began to enter college, record numbers of students were enrolling in mathematics and science courses (Johnson, Johnson, & Smith, 1991). To meet the increased demand, universities and colleges began offering these courses in large lecture halls with one professor lecturing to several hundred students at a time. The size of these classes precluded personal interaction between faculty and students and led to less interaction among students; the task of testing and grading this number of students led to more standard, routine tests (Teles, 1992). This may partly explain the dramatic drop in students' interest in math and science. The percentage of first-year students interested in undergraduate mathematics majors dropped from 4.6% in 1966 to 0.6% in 1988 (Green, 1989), and an estimated 50% of students who chose science, mathematics, and engineering majors when they entered college switched to non-science majors, mostly within their first two years of college (Green, 1989; Hilton & Lee, 1988).

In an attempt to identify why undergraduate students leave science, math, and engineering (SME), Seymour and Hewitt asked SME seniors and former SME majors about their educational experiences (Hewitt & Seymour, 1992; Seymour, 1995; Seymour & Hewitt, 1997). One of their major findings was that switchers and non-switchers were not different types of students: they were similar in their abilities, attitudes, high school preparation, willingness to work hard, and even in their grades in SME courses. A primary difference between the two groups of students was that the non-switchers used a variety of coping strategies and resources to help them surmount the difficulties they faced. For example, those who stayed in SME majors were more likely to have worked with peer study groups to gain a better understanding of complex material. Small group learning is one of the factors they cited as having helped them to survive, and some switchers considered this omission to have contributed to their leaving.

Reports on faculty perceptions of the high rate of student attrition discuss such causes as inadequate facilities, large classes, and poor pre-college preparation of students (National Advisory Group, 1989; National Science Foundation, 1988). As these concerns became more widespread, more college instructors experimented with alternative ways to teach mathematics. This set the stage for a major movement from within the mathematics community to reform the teaching of calculus (Douglas, 1986), a course that serves as a gateway to most science and engineering fields. While many mathematics instructors had been experimenting with innovations in calculus instruction for some time (for example, see Teles, 1992; Treisman, 1985), the Tulane Conference in 1986 marked the beginning of an organized and recognized movement within the mathematics community (Douglas, 1986).

Group work was used by pioneers of calculus reform long before 1986 (for example, Davidson, 1971; Teles, 1992; Treisman, 1985), but the power of cooperative learning to catalyze students' conceptual understanding of difficult calculus concepts was underestimated early in the movement (Schoenfeld, 1995). Around 1988, reformers began to shift more of their attention from curriculum issues toward issues of pedagogy. At some levels, this change was dramatic. There are indications, for example, that the NSF made a change in a key criterion for funding calculus reform proposals (based, presumably, on a shift in reviewer opinion) from one year to the next in which a focus on pedagogy changed from a very negative to a very positive attribute of a proposed project (Ed Dubinsky, personal communication, August 28, 2000). This new focus on pedagogy included a renewed interest in cooperative group work. For some instructors, having students work in groups was initially a byproduct of insufficient resources, such as when students were forced to share computers or when team homework was assigned to decrease instructors' workloads (Brown, 1996; Hagelgans et al., 1995; Ross, 1996). Although the extent, nature, and quality of calculus reform efforts are difficult to document overall, by early 1995 over two-thirds of the mathematics departments surveyed indicated that some reform efforts were underway at their institutions (Leitzel, 1995), most of which included some component of group work (see for example Leitzel, 1995; Tucker & Leitzel, 1995).

As group work became part of college mathematics instruction through these various avenues, instructors and researchers have used the terms *cooperative learning*, *group work*, and *collaborative learning* to refer to a variety of instructional strategies. While we use these terms interchangeably, it is important to recognize that some authors have distinct definitions of these terms, and that they collectively refer to a broad range of classroom activities. Putting students in groups to work together is generally recognized as insufficient to realize the benefits of cooperative learning—learning tasks need to be structured to ensure interaction among students, and classroom norms need to be established that support that interaction (Davidson, 1990; Hagelgans et al., 1995; Johnson & Johnson, 1994b). The literature offers conflicting results on the effectiveness of cooperative learning methods; these apparent conflicts may arise from the different ways in which cooperative learning is implemented (Cohen, 1994). These various models of cooperative learning are an important factor in interpreting research and other reports about cooperative learning in calculus.

With its increasingly widespread use, what do we know about effects of cooperative learning in calculus? Most information about the efficacy of reforms appears as anecdotal reports (see, for example, Alvarez, Finston, Gehrke, & Morandi, 1993; Armstrong, Garner, & Wynn, 1994; Hilbert, Maceli, Robinson, Schwartz, & Seltzer, 1993; Solow, 1991; Tucker, 1995, 1990). Many writers about calculus reform have acknowledged that the role of this material has been to disseminate information, rather than to research the effects of the reforms on students. As one reformer put it, "We always saw course development itself as our first priority" (Ostebee & Zorn, 1990, p. 207). The role that educational research should play in guiding reform efforts is generally not discussed. As a result, there is little research-based evidence that allows evaluation of the reforms.

Cooperative Learning in Calculus. In contrast to the lack of research at the post-secondary level, there is extensive research on cooperative learning methods and their efficacy at the primary and secondary levels. This research has established that cooperative learning can be effective in promoting achievement, motivation, creativity, self-esteem, and positive relationships among elementary students (see Johnson & Johnson, 1994b; Slavin, 1995). The results have been more mixed for secondary mathematics (for example, Davidson, 1985; Newmann & Thompson, 1987; Owens, 1995). The range of ways cooperative learning is used in post-secondary calculus classrooms is not well documented, nor is research on its effects. In this section, we review research on college calculus courses that included a focus on cooperative learning.

A meta-analysis on the effects of cooperative learning in post-secondary SME courses found that cooperative learning methods generally do improve academic achievement (Springer, Stanne, & Donovan, 1997). This analysis included studies from a broad range of courses, more than 50% of which (13 out of 23) were mathematics courses. Of those, one was about computer science, three or more looked at complex classroom changes of which cooperative learning was but one component, and several looked at remedial mathematics or methods courses for pre-service teachers. Only three of those papers tried to compare cooperative learning with teacher-centered learning, and only one addressed calculus.

Duke University was among the first institutions to receive NSF funding for calculus reform. Begun in 1988, Project CALC (Calculus As a Laboratory Course) featured "real-world problems, hands-on activities, discovery learning, writing and revision of writing, teamwork, and intelligent use of available tools" (Bookman & Blake, 1996). In particular, a weekly lab was added to the course in which students worked in pairs, using computers to discover and investigate complicated calculus concepts. In addition, the three regular weekly class meetings were restructured from being mostly lecture to including a significant amount of team work, where students worked on substantial problems and summarized their results in biweekly written reports (Smith & Moore, 1990).

In a comprehensive evaluation of Project CALC (PC), PC students did significantly better on a problem-solving test than traditional (TR) students, but did slightly worse at computational skills (Bookman & Friedman, 1994). As a result, Project CALC was modified to include more computational homework and different exams. Students initially reported a great deal of uneasiness, preferring the sense of mastery that came with straightforward computational work; the new course forced them to question deeply held beliefs about mathematics. However, after a semester, most students began to accept the underlying philosophy of the reforms, and their initial negative reactions softened. Qualitative data indicate that PC students were more actively engaged during class time than TR students were. The first large group of PC students performed slightly less well (0.2 grade points per course) in higher-level mathematics courses than TR students. Even though Project CALC included a strong component of group work, the assessment was designed to measure the overall effects of the program, so the specific effects of group work are unknown.

At Purdue University, Ed Dubinsky and Keith Schwingendorf developed Calculus, Concepts, and Computers with Cooperative Learning (C^4L), a calculus reform effort grounded in constructivist perspectives on how mathematics is learned (Dubinsky, 1992). With the goal of helping "students construct appropriate processes and objects and get them to reflect on and use these constructions" (p. 48) in calculus, they developed their own course materials, including a text, homework assignments and laboratory activities (p. 48). They altered the three lectures per week to involve significant use of group work and moved the two weekly discussion sections into a computer lab. Unlike with Project CALC, the computer labs concentrated on using a mathematical programming language (ISETL) to write programs that modeled various calculus concepts.

During the first year of the program, C^4L students took the same final exam as students in traditional sections. The last portion of each semester was spent preparing students for this (traditional) assessment; the C^4L students performed as well or better than their peers on this exam, and received slightly higher course grades than their counterparts. C^4L students had higher retention rates in the calculus sequence and higher overall GPAs (at their last available reporting); the two groups took roughly the same number of mathematics courses beyond calculus (Schwingendorf, McCabe, & Kuhn, 2000; Dubinsky & Schwingendorf, 1997; Dubinsky, 1992). A detailed description of the C^4L program and its results is given in (Schwingendorf et al., 2000).

Vidakovic interviewed C^4L students about their understanding of inverse functions; some were interviewed as a group, others alone (Vidakovic, 1997). She concluded that a student's group played a key role in creating cognitive conflict, or disequilibrium, the Piagetian stage preceding cognitive development. Working alone, students were much less likely to see the contradictions inherent in their thinking, and more likely to ignore them when they were seen. Although the interview setting differs considerably from that of the classroom, her findings shed light on a mechanism through which group work might affect student learning.

At the University of Michigan (UM), Morton Brown and others developed Michigan Calculus (Brown, 1996), which included the use of graphing calculators, a reformed textbook (Hughes-Hallet & Gleason, 1992), team homework, cooperative learning in the classroom, and a training program for TAs and instructors. Among the listed program goals were students becoming better life-long learners by working cooperatively, and improving instruction with a student-centered approach and instructor-facilitated cooperative learning (Burkam, 1994b).

Much of the data about Michigan Calculus covers the transitional period when the reforms were changing from the limited scope of a few classes to taking over all calculus sections. On a test of conceptual calculus problems, students from Michigan Calculus performed significantly better than students in traditional classes. They concluded that students in the reform program had "an enhanced geometric understanding of functional behavior and an improved ability to work with mathematical models" (Burkam, 1995, p. 7). The strongest effect of the reforms came in the area of student attitudes and beliefs. Compared with students in traditional discussions, students in the reformed classes "perceived a greater relevance of their mathematics coursework to other classes, post-college activities, and to 'real world' problems. Perhaps most importantly, they expressed a deeper interest in mathematics and a greater confidence in their own mathematical abilities" (Burkam, 1994a, p. 7). Effects on persistence were less clear. Over the first two years, students in the reform sections of Calculus I were more likely to continue on to the second semester of calculus. However, this effect disappeared when the higher course grades of students in reform sections were taken into account. As with Duke's Project CALC, it is not clear which (if any) of Michigan Calculus's effects were related to their use of cooperative learning in the classroom.

Armstrong, Garner, and Wynn (1994) compared two different reform programs (one based on Hughes-Hallet & Gleason, 1992 and the other based on Stroyan, 1993) with a traditional class. Both reformed classes included cooperative learning. Based on teaching evaluations and student journals, they found that both reform classes drew more positive comments than the traditional classes. Analysis of students' grades in subsequent classes for which calculus was a prerequisite showed no significant differences among the three treatments.

Solow (1991) and Hilbert, Maceli, Robinson, Schwartz, and Seltzer (1993) introduced group projects and some degree of discovery learning into their calculus classes. Both studies used qualitative data (including student and instructor comments) to argue that their reformed classes significantly increased student engagement in calculus. Hilbert et al. also argued that the reformed classes gave students a broader view of mathematics as a creative endeavor.

Concerned about the failure of many African-American students at the University of California at Berkeley to successfully complete the calculus sequence, Treisman compared their study habits with those of Chinese-American students who had been far more successful in navigating these gateway courses (Fullilove & Treisman, 1990; Treisman, 1985). He found that the African-American students studied mostly alone, while the Chinese-American students spent a substantial amount of time working together in groups. He judged this peer support structure to be integral to their success in calculus classes, and developed a program called the Professional Development Program, later called the Emerging Scholars Program (ESP). The program was known as an honors program in which students, primarily minorities, attended traditional lectures, took exams with the other students, but met in special workshops for four hours every week. In the workshops, they worked in cooperative groups on challenging calculus problems assisted by both an instructor and an undergraduate student assistant. In addition, they participated in social events intended to build a sense of community. Similar programs have been implemented at many institutions across the country, including the University of Texas at Austin, California State Polytechnic University at Pomona (Cal Poly), and the University of Wisconsin at Madison.

Compared with non-ESP students, students who participated in these cooperative learning-based ESPs at various institutions excelled in a variety of ways, including higher average grades, increased persistence in mathematics-based majors, and higher graduation rates (Fullilove & Treisman, 1990; Alexander, Burda, & Hwang, 1995). At UT-Austin, ESP also significantly increased the number of minority math majors (Uri Treisman, personal communication, March 3, 1995). At Cal Poly, ESP students completed the three-quarter calculus sequence an average of one quarter sooner than the non-ESP students (Bonsangue, 1994; Bonsangue & Drew, 1995). At UW-Madison, classroom observations and student interviews in the second year of the program indicated that ESP students developed a community, built communication skills, learned problem-solving skills, gained self-confidence, and were inspired to learn more mathematics (Alexander et al., 1995).

In an attempt to generalize the ESP model to mainstream calculus students, we describe below a study which compares traditionally-taught discussion sections, in which the TA worked homework problems at the blackboard, with sections using cooperative group work modeled after ESP workshops.

Some Limitations of This Research. These studies exhibit a number of problems, which limit the conclusions about cooperative learning that can be drawn from them. In particular, these studies exhibit problems of sample selection bias, non-independence among the units of analysis, and lack of experimental controls to isolate the effects of explanatory variables.

Sample Selection Bias. The students in Emerging Scholars Programs participated by application or invitation, and so they represent a specially-selected sample. It is unknown whether similar success would be realized with a randomly-selected sample. For example, at Cal Poly, each African-American and Latina/o student who was admitted into the College of Science or Engineering was invited to participate in the ESP program; about half of the students accepted. Similar enrollment procedures have been used at other ESPs. Students who agreed to participate in any of the ESP programs were volunteering to spend substantially more time in calculus class each week. Because of the volunteer nature of student participation, it is possible that the significant improvements in grades and retention for the workshop students reflect other characteristics of the students who chose to participate.

In educational research in most college settings, the way students sign up for classes makes a truly random sample of students a practical impossibility. Researchers have tried to account for this in several ways. One approach is to compare background characteristics of students in the different treatments, and to incorporate those comparisons into interpretation of the results. At UC-Berkeley, Treisman noted that workshop students who graduated or stayed enrolled had lower SAT-Math scores than the non-workshop persisters (Fullilove & Treisman, 1990). In Duke's Project CALC and in the ESP program at UT-Austin, no significant differences were found in either Math or Verbal SAT scores between the students in the reformed and traditional classes (Bookman & Friedman, 1994; Myers & McCaffrey, 1994). A more accurate approach is to incorporate differences in background variables in statistical analyses, using analysis of covariance or other methods. In studying the ESP program at the UW-Madison, Levin considered gender, ethnicity, high school GPA and rank in class, ACT scores, mathematics placement test scores, and other variables, based on four prior cohorts of students, as potential covariates (Levin, 1995). Based on a finding that high school percentile rank in class and placement test scores were the best set of predictors of grades in Calculus I and II, he used these variables as covariates in the analysis of course grades. A similar method was used to study the C^4L program (Schwingendorf et al., 2000).

In most of these studies, similar adjustments or comparisons could not be made for possible differences in students' initial attitudes or motivation, since no measures of these characteristics were available. Although statistical adjustment by analysis of covariance or other comparisons does not substitute for true random sampling, often they are the best that can be done.

Units of Analysis. Many of these studies suffer from another pervasive problem in classroom research, referred to as the "units of analysis" problem (Levin, 1995). In studies like the ones described here, each classroom was associated with its own unique instructional experience, which differed in various ways from one classroom to the next. In these cases, the statistical assumption of inter-student independence is clearly violated, and it is therefore not statistically valid to include individual students as independent units in the data analysis. Instead, entire classrooms constitute the independent units of investigation, and more appropriate statistical analyses should be based on comparisons among classrooms aggregated over groups of students.

In studies in which large numbers of classrooms are being compared, this non-independence problem can be addressed by the use of hierarchical or nested linear models, or other statistical methods that directly compare classrooms, rather than students. In the analysis of achievement data from the ESP program at UW-Madison, covariate-adjusted grades were first averaged for all students in each classroom section, and the classroom averages were then compared (Levin, 1995). In our study of cooperative learning described below, we use the statistical technique of nested linear models, in which students were "nested" within discussion sections, and the classroom averages (after adjusting for covariates) formed the basis of the treatment comparisons.

In studies comparing small numbers of classrooms, these methods might not be practical. If statistical comparisons are made directly among the students in different classrooms, then standard errors may be underestimated and significance tests may be overly liberal. In other words, when comparing students who experienced different classroom treatments, significant findings might be due to the treatments or to any of the other myriad ways in which the students' or their classroom experiences differed. This confounding must be considered in interpreting research results.

Controlling Variables. Most reform projects that included some form of controls varied a large number of factors between the reform and traditional classes. By changing so many things at once, adequate controls are often not available to isolate the effects of any particular change. In Duke's Project CALC, Purdue's C^4L, and Michigan Calculus, the course content, pedagogy (including the use of calculators and computers), and most exams were different. ESP participants spent more time in class, worked cooperatively in small groups in class, received more concentrated support and attention from faculty and other instructional staff, worked on challenging, conceptual problems, and participated in social events together. On the basis of the evidence available, it's not possible to identify which of these elements create which of the observed effects, or if some of them are more influential than others.

The studies described above have shown that cooperative learning-based calculus classes can lead to a variety of positive effects, including increased grades, improved conceptual understanding, more positive attitudes, and increased retention in calculus and in mathematics-based majors. However, we do not know whether the positive effects result from cooperative learning, or from the other changes in the curriculum. These effects do come at the expense of significant investments of resources, often requiring substantial commitments of faculty time, greater class time, new classroom materials, and computers and calculators. If a mathematics department wants to implement those changes which will be most effective for its population of students, in a cost-effective manner, then they need to know which types of reforms produce which types of effects, a question which has not yet been adequately addressed.

A Study of Cooperative Learning in Calculus

The idea of controlling variables was a motivating force behind our study of cooperative discussion sections. We attempted to isolate the effects of cooperative learning by keeping all other conditions as similar as possible between the two treatments, despite our interest in other changes, such as the use of technology or student projects. Once we understand the effects of these initial changes, then further reforms can be considered, and studied, in the future.

The model for the experimental course came from the ESP at UW-Madison. As is true at many universities, the large-lecture format currently used for calculus classes was failing to reach many of the students. Since Emerging Scholars Programs have generally been found to be associated with higher course grades, better attitudes, and lower attrition within small groups of specially-selected students, this study was designed to examine whether a similar approach could be used successfully within the general population of Calculus I students.

The focus of the study is collaborative group work within discussion sections, compared with the more traditional teacher-centered discussion sections. To allow a direct, controlled investigation of the effects of cooperative learning, we chose to keep the course as similar as possible to its usual format, and resisted the temptation to implement other changes, such as the use of technology or an emphasis on applied projects and student writing.

Course Format. Students in our Calculus I course typically spend two and a half hours in lecture per week, with about 150 to 250 students enrolled in each lecture. They also attend two 50-minute discussion sections of about 22 students, led by a graduate student teaching assistant (TA). The basis of the experimental course was to change the format of the discussions, from TAs answering questions and working homework problems at the blackboard to students working in small groups to solve challenging problems. Following the ESP model, the problems were intended to be thought provoking, requiring collaboration to complete, and building both basic skills and conceptual understanding of the material.

ESP students at UW-Madison meet in their discussion groups for six hours per week, for which they earn additional academic credit. Some critics might argue that the effectiveness of ESP has been due to the longer time students spend in their discussions. Since the standard 50-minute discussions may be too short to be productive for cooperative groups, we included a comparison of short (50-minute) and long (75-minute) discussions, each meeting twice per week, to investigate the effect of the length of the discussions.

An experienced senior professor taught the course twice, in back-to-back lectures meeting for 50 minutes three times per week, attempting to keep the lectures as similar as possible. Seven TAs taught two discussion sections each, following the two-way design shown in Table 1.

Teaching Assistants. Each TA led two discussions, both of the same type and length. The differences among the actual classrooms—encompassing factors such as the TA herself or himself, the time of day, and the physical classroom setting—each contribute to the students' experiences. Since we had a relatively small number of classes in each of the four treatments, it was especially important that the TAs leading those sections be as comparable as possible in teaching skill and experience. To achieve this, TAs were assigned to treatments by matched assignment; one relatively inexperienced and one experienced TA were assigned to each treatment condition (except for short, traditional, in which there was only one TA because of lower than expected enrollment; see Table 1).

All of the experienced TAs had several years of teaching experience and had won awards for their teaching (the authors were among the experienced TAs). The novice TAs were in their first or second year as TAs. To further equalize their skills, all TAs participated in a training program before the start of classes. The TAs teaching cooperative sections also participated in a training program specially designed to address cooperative learning.

Table 1. Study Design and TA Assignment. Each TA taught two discussions of the same length and format.

		Discussion Format	
		Cooperative	Traditional
Length of discussion	Short (50 minutes)	4 discussions 1 experienced TA 1 novice TA	2 discussions 1 experienced TA
	Long (75 minutes)	4 discussions 1 experienced TA 1 novice TA	4 discussions 1 experienced TA 1 novice TA

Discussion Sections. Although we labeled the discussion sections "cooperative" and "traditional," individual TAs' teaching styles influenced what actually took place in their classrooms. To evaluate the extent to which each classroom actually fit into its category, the TAs kept daily records of what happened in their classes.

In the traditional sections, all the TAs spent the vast majority of class time (between 79% and 95%) at the board, doing problems and answering students' questions. The remainder of the time was spent giving quizzes, taking care of administrative details, and other miscellaneous activities.

The four TAs teaching cooperative sections worked together to produce ESP-style worksheets for every class period. Group work during a discussion section meeting consisted of the TAs placing students into groups of three or four to work on the worksheets. In the 50-minute discussions, the majority of class time was spent doing group work (79% of the time for the experienced TA, 67% for the novice). These TAs spent relatively little time at the board (18% and 31% respectively). Interviews with these TAs indicated that although their students wanted them to spend more time at the board, they felt pressured by the lack of time and the need to give students sufficient time to discuss the worksheet problems. In the 75-minute discussions, the TAs split their time more evenly between teaching from the board and having small groups work on the worksheets. The experienced TA spent 64% of her time at the board and 34% doing group work; the novice TA's split was 42% at the board and 57% group work.

It should be noted that because of the longer class period, the smaller percentages of group work time in the longer discussions do not necessarily indicate less time spent in groups (e.g. 75% of a 50-minute class is the same as 50% of a 75-minute class). Thus in all the cooperative sections (short and long), an average of between 30 and 40 minutes per class period was spent doing group work. Because of the extra 25 minutes, TAs teaching longer sections were able to spend more time working at the board.

Students. In an ideal experiment, students would be randomly assigned to each of the four treatment conditions, possibly after matching them on characteristics such as prior mathematics grades or placement test scores. In this way, the four groups of students could be considered to be random samples, and therefore comparable. However, as at most institutions, our students register for any open section of calculus they choose. This raises a concern about how the students decided which lecture and discussion section to register for. It is possible that the different lengths of the discussions (apparent from the course timetable used in registration), and the time of day the lectures and discussions are offered, might have produced a registration bias. We attempted to minimize this bias by avoiding drawing attention to the differences between the two lecture courses.

On a survey at the beginning of the semester (described in more detail below), students were asked why they chose the section they did. Seventy-eight percent chose the response "It was the section that fits my schedule best." Seven percent indicated that they preferred a particular length discussion (3.5% indicating preference for each of the 50-minute and 75-minute discussions). Three percent chose their section based on the reputation of the professor or to be with their friends.

Students also were asked for demographic characteristics such as gender, race and/or ethnic group, and their prior experience with calculus classes. From registrar's data, we collected information on SAT and ACT scores and scores on mathematics placement tests. These factors were considered as covariates in the statistical analysis.

Outcome Measures. We evaluated student outcomes in terms of achievement, attitudes about mathematics and perceptions about themselves as learners, and persistence in the calculus sequence.

Student Achievement. Evaluation of student achievement was based on scores on six exams given throughout the semester. All exams after the first consisted of open-ended problems (the first was multiple choice). To ensure uniformity of grading, each TA and the lecturer graded a specific question (or set of questions) for all students enrolled in both lectures. The lecturer designed the exams with some input from the TAs. Exam content was not modified in any way to accommodate the new course format; that is, the exams comprised a test of how students performed on traditional measures of achievement. All exam scores are reported as percentages.

Student Attitudes. Students in both lectures completed an attitude survey at the beginning and end of the semester. Administering the survey twice allowed us to look at changes within individuals, and also to partially control for possible baseline differences among the students in the four treatment conditions. To assure maximum participation, the surveys were administered in lecture, on the same days as the first and fifth in-class exams.

The survey was adapted from the Fennema–Sherman Mathematics Attitudes Scales (MAS) (Fennema & Sherman, 1986). From the eight dimensions of those scales, we included four: Confidence in Learning Mathematics, Attitudes Toward Success in Mathematics, Mathematics Usefulness, and Effectance Motivation in Mathematics. Two additional scales were constructed for this study: Beliefs about Mathematics, and Learning with Others, following a similar format to the MAS scales. Many of these items were adapted from the University of Michigan and other sources (Burkam, 1994a; Kloosterman & Stage, 1992; Schoenfeld, 1989) with others developed to balance the type and distribution of items. Each scale consists of 6 positively-worded and 6 negatively-worded items. More detailed definitions and a list of the items in each of the six scales are shown in the Appendix.

The 72 scale items were arranged in a pseudo-random order in which no two subsequent items were from the same scale, and positive and negative items alternated. Respondents indicated the degree to which they agreed or disagreed with each item on a 5-point Likert-type scale. Positively-worded items were assigned scores ranging from $+2$ for a response of "Strongly Agree" to -2 for "Strongly Disagree." For negatively-worded items, scoring was reversed. Scores were then averaged over the 12 items of each scale to produce the scale values, so that values on these scales could range from a possible low of -2.0 (e.g. low confidence) to a possible high of $+2.0$ (e.g. high confidence).

The analysis of attitude scales was based on the change in scale values from the beginning (first survey) to the end (second survey) of the semester. Positive differences indicate increases (e.g. increased confidence).

Persistence. For all students who completed the course, we used registrar's data to determine whether they continued on to Calculus II in the following semester or the subsequent summer. We also compared their course grades in Calculus II. However, we do not know what types of classroom experiences these students had for Calculus II. Their grades in the second course are more likely to reflect their experiences in that course than in the experimental course described here.

Data Analysis. To avoid the units of analysis problem (Levin, 1995), our analysis was based on comparisons among the TAs' discussion sections, rather than among the individual students enrolled in them. Results are based on a nested linear model, with section means nested within TAs, which are then nested within treatments. Type and Length were included as crossed factors, and an interaction term was included. The TAs' level of experience (experienced or novice) was also included as a crossed factor in the design. Because of the small number of TAs, we were not able to investigate possible interactions between experience and the other design variables.

Means were adjusted (by analysis of covariance) for students' quantitative ACT scores,[1] mathematics placement test scores, and attitude scale values at the first survey. For all exams after the first and for the attitude variables, the score on exam 1 was also used as a covariate.

Results

Of the 313 students who began the course, 27 dropped the course by the end of the semester. The proportions of students who dropped from Cooperative or Traditional sections did not differ significantly ($\frac{20}{181}$, or 11.0% from Cooperative sections and $\frac{7}{132}$, or 5.3% from Traditional sections).

Based on responses to the first survey ($n = 313$), the six scales all showed at least reasonable reliability. For the MAS scales, Cronbach's alpha ranged from 0.86 to 0.92. The two scales developed for this study had alpha values of 0.69 (Beliefs about Mathematics) and 0.86 (Learning with Others).

The average value on each of the scales was slightly positive at the beginning of the semester, with Attitudes Toward Success and Usefulness having means greater than 1.0 (on a scale of −2.0 to +2.0). Over the course of the semester, mean values on five of the six attitude scales dropped significantly. Attitudes Toward Success had the highest initial mean, and also changed the least over the course of the semester. The remaining five scales dropped by an average of 20% to almost 50% of a standard deviation ($p < 0.001$; see Table 2).

Table 2. Mean Scores on Attitude Scales

Scale	Survey 1	Survey 2	Effect Size
Confidence in Learning	0.81 [a]	0.58 [b]	−0.40 [c]
Mathematics	(0.036)	(0.041)	
Attitudes Toward Success	1.17	1.15	−0.05
in Mathematics	(0.030)	(0.031)	
Mathematics	1.13	0.87 [b]	−0.49
Usefulness	(0.031)	(0.037)	
Effectance Motivation	0.60	0.38 [b]	−0.41
in Mathematics	(0.031)	(0.037)	
Beliefs About	0.84	0.76 [b]	−0.20
Mathematics	(0.023)	(0.029)	
Learning With	0.22	0.06 [b]	−0.30
Others	(0.032)	(0.53)	

a. Mean (sem). Includes all students who completed both surveys ($n = 284$). Values on these scales could range from a possible low of −2.0 (e.g. low confidence) to a high of +2.0 (e.g. high confidence).

b. Significantly different from Survey 1 mean, $p < 0.001$.

c. (Survey 2 mean − Survey 1 mean) / (Survey 1 standard deviation)

Students who had previously taken calculus in high school, but not in college, scored significantly higher on exams than students who had not taken calculus before (the difference was between 10 and 14 percentage points for all exams except exam 4; $p < 0.001$). Students who had taken calculus before in college fell between these two groups, on average. This finding would introduce a bias in favor of those

[1]ACT scores were not available for all students. For students who took the SAT instead, ACT scores were assigned corresponding to the same percentile score as the student's SAT score (Marco, Abdel-fattah, & Baron, 1992).

sections with higher proportions of students who had taken calculus in high school. In order to make the sections more comparable to each other in this variable, we restricted further analysis to include only those students who had taken calculus before in high school, but for whom this was their first college calculus course. Of these, only those students who completed the course, took all 6 exams, and had values for all covariates (ACT, placement tests) were included.

The analysis was based on 189 students (66.1% of all the students who completed the course, with between 8 and 17 in each section), of which 98 were in Cooperative sections and 91 were in Traditional sections. Once the sample was limited in this way, the distribution of gender across the sections was quite variable (ranging from 45% to 100% male), which precluded us from examining gender effects in this analysis. Similarly, we were unable to examine the effects of race/ethnicity, since there were very small numbers of non-white students in each of the sections.

Only placement test scores and scores on Exam 1 were significant covariates for exam scores. For change in attitudes between the two surveys, only the corresponding scale values at the first survey were significant. The adjusted means reported below reflect the model that includes all the covariates; similar results were obtained if the covariates were omitted from the analysis.

Adjusted mean exam scores, both overall and within treatments, are shown in Table 3. For most exams, the mean score was approximately 71–74%, except for exam 4, for which the mean was much lower. There were no significant treatment differences or interactions for any of the exams. All treatments had almost identical means on the weighted mean of the exams.

Adjusted mean scores for changes in attitude scales between the first and second survey are shown in Table 4. There were no significant treatment differences or interactions for most of the scales. The only significant findings were a length effect ($p < 0.009$) and a length by type interaction ($p = 0.027$) for Effectance Motivation. However, given the large number of significance tests we conducted, it is likely that these results are spurious.

Table 3. Adjusted Exam Means Overall and by Treatment

Treatment	Exam 1	Exam 2	Exam 3	Exam 4	Exam 5	Final	Weighted Mean [a]
Overall	73.4 [b]	72.1	72.2	53.9	70.5	73.9	68.9
Cooperative	71.7	74.7	71.1	54.1	72.6	73.2	69.2
Long	70.8	75.1	70.8	54.6	69.7	74.3	69.4
Short	72.6	74.4	71.4	53.5	75.5	72.2	69.0
Traditional	75.1	69.4	73.3	53.7	68.5	74.6	68.6
Long	72.4	70.0	70.1	54.6	71.1	72.6	67.9
Short	77.7	68.8	76.6	52.7	65.9	76.7	69.3
Standard Error [c]	0.3	4.0	2.6	6.7	7.2	3.2	3.1

a. Weighted mean of exam scores; weights were determined by the lecturer during the semester for calculating course grades (relative weights of 1, 3, 3, 3, 1, and 4 for the five exams and the final, respectively).

b. All exam scores were converted into percentages for this analysis.

c. Square root of the Mean Square Error (2 df) from the analysis of covariance. Because of the non-independence of students, these underestimate the true MSE.

Table 4. Adjusted Means Overall and by Treatment for Changes in Attitude Scales

Treatment	Confidence in Learning Math	Attitudes Toward Success	Math Usefulness	Effectance Motivation in Math	Beliefs About Math	Learning with Others
Overall	−.25 [a]	.01	−.30	−.20	−.12	−.17
Cooperative	−.24	−.03	−.31	−.20	−.12	−.17
Long	−.26	.03	−.32	−.23	−.15	−.19
Short	−.23	−.09	−.30	−.16	−.10	−.16
Traditional	−.26	.06	−.29	−.20	−.11	−.17
Long	−.27	−.05	−.32	−.31	−.06	−.17
Short	−.25	.16	−.25	−.08	−.16	−.17
Standard Error [b]	.10	.06	.03	.04	.05	.21

a. Values on the scales could range from a low of −2.0 (e.g., low confidence) to a high of +2.0 (e.g., high confidence). The analysis reported here was based on the change in scale values from the beginning (first survey) to the end (second survey) of the semester. Positive differences indicate increases (e.g., increased confidence).

b. Square root of the Mean Square Error (2 df) from the analysis of covariance. Because of the non-independence of students, these underestimate the true MSE.

We originally included TA experience as a factor in the design just as a statistical control; since the design was not balanced for this variable, more detailed analyses based on experience were not possible. However, there were a number of interesting differences based on TA experience that might warrant further research. For all five of the classroom exams, sections taught by experienced TAs had higher mean exam scores than sections taught by novice TAs, with differences ranging from less than one percentage point on Exam 1 to 10 points on Exam 5. On the final, sections taught by novice TAs had slightly higher means than sections taught by experienced TAs. Although these findings were not significant given our small sample of TAs, they are suggestive.

Of the 286 students who completed this course, 163 (57.0%) completed Calculus II in the following semester or the subsequent summer. There were no treatment differences or interactions in the proportion of students who completed Calculus II after this course. Of the students who completed Calculus II, those who had enrolled in short discussion sections of Calculus I had higher grades than students who had enrolled in long sections.

Discussion

One of the motivating questions of this research was "Can the ESP model be generalized successfully to the mainstream calculus population?" Even with an experienced ESP instructor, using similar types of student tasks as in ESP, our pseudo-random selection of students in a calculus discussion did not achieve scores on exams as high as we are accustomed to seeing in ESP (Millar, Alexander, Lewis, & Levin, 1995). However, the lack of significant treatment differences in this study is a significant finding. The conventional wisdom is that cooperative learning requires more class time, and that as a result, less material can be covered, to the students' detriment. Contrary to this idea, the short cooperative sections did not do any worse than either of the traditional treatments. The TAs in the cooperative sections generally enjoyed teaching their classes that way (one commented, "I finally see what the students are really thinking!"). This finding gives instructors solid evidence that this method of teaching, even within the same amount of class time, need not hurt their students' performance.

There are several possible reasons why the success of ESP did not occur for our more general population. In addition to the effects of a reformed course, a particular course or program might have very different effects on different groups of students. In an extension of the ESP workshop model to a pre-calculus course, Ganter (1994) explained the lack of treatment differences in part by the argument that the students at UC-Berkeley (where the initial successes of ESP were reported) had stronger academic preparation than the students in the community-college setting where she conducted her study. "Stronger academic training may mean that the students are better prepared to work independently and that they are more likely to have the confidence necessary to initiate the exploration and learning of mathematical ideas in a study group setting without instruction from the workshop leader" (p. 13). In the present study, our students were a different population than the ESP students, which may have affected their confidence, motivation, or other aspects of their preparation and skills. In post-workshop interviews, Ganter found that "most of the workshop students stated that [the] lack of community at a 'community' college made it almost impossible for them to form peer support groups, a vital part of the workshop program" (p. 14). Thus the program (ESP) that has been shown to be so effective in a specially-selected group of students simply may not be successful in the broader student population.

Further, some studies have shown that female students are more successful in classrooms that promote cooperative activity and student engagement with mathematics (Boaler, 1997; Peterson & Fennema, 1985; Rogers, 1990). Given recent concerns about increasing the recruitment and retention of women and minorities in mathematics, investigating the effects of these reforms on particular populations of students is critical. The distribution and proportions of women and minorities in the discussion sections in our study prevented us from investigating these questions with these analyses. Gender differences will be examined in a future paper.

The length of the discussions is another factor at play in ESP. One common belief about ESP is that it succeeds simply because of the extended time that students spend in discussion. In the controlled setting of this experiment, this belief was not upheld; long sections did not score differently from short sections, either within cooperative or traditional instruction. However, we only compared relatively short discussion sections; if length of time spent in discussion is an important variable, then its impact occurs at more than the two and a half hours per week that we investigated.

This leads to the interpretation that the high exam scores achieved by ESP students do not result only from the length of the discussions, nor solely from the specific cooperative learning model used. It is likely that the success of ESP results from some combination of the various aspects of the program, including length of time, use of group learning, types of students, and community-building activities. Also, students in ESP are periodically encouraged to reflect on the group learning process, an activity which is advocated as important for the effectiveness of cooperative learning (Cohen, 1994; Johnson, Johnson, & Holubec, 1991; Johnson et al., 1991). Because of the limited time spent in discussion in the current study, we did little to encourage this type of reflection. Thus the answer to the question "Would the ESP model be successful?" is "No, not necessarily." "Could the ESP model be successful?" is still unanswered; more targeted research would need to be conducted to isolate the effects of other aspects of the program. For example, one might test the ESP model without using cooperative learning, but keeping all other aspects of the program (6 hours of workshops per week, for example) the same.

Although the drops in attitude scales across the semester are relatively small, the attitudes that they reflect have developed in students over the 12 years or more that they have been learners of mathematics. We would not expect them to change substantially over the course of only three months. Given this perspective, the magnitude of the changes we saw are compelling. They provide evidence that the experience of first-semester calculus has a dramatic effect on student attitudes, resulting in more negative attitudes toward mathematics and toward themselves as learners of mathematics. The second survey was administered three weeks after Exam 4, on which the scores were very low, and it is possible that the large decreases we saw were a result of the low scores on that exam. However, it is interesting to note that one scale, Attitudes Toward Success, did not change, despite this timing.

This study was not designed to investigate the role of TA experience. Although the main effects for TA experience were not significant, they were consistent, and suggest that this might be an interesting target for further study. Effective use of cooperative learning requires a lot more than just assigning students to do work in groups (Cohen, 1994; Johnson et al., 1991; Slavin, 1995). Davidson discusses the "basic management and leadership functions that must be performed—many of them by the teacher" in small-group instruction (Davidson, 1985, p. 211). For teachers using cooperative learning in their classrooms, it might not be years of teaching experience that matters, but the amount of specific training and experience they have in facilitating group work. Alternatively, it may also be that TA experience interacts with the type of pedagogy used; years of TA experience might matter less in courses that emphasize group learning since in that setting, students learn more from each other than directly from the TA. In teacher-centered classrooms, TA experience might have a greater impact than in those based on cooperative learning.

The measures of student achievement play a key role. In this study, and in most of the studies discussed earlier, students in reformed courses were given the same types of exams that are given in traditional calculus courses. Burkam reports on an attempt to "construct a more authentic measure of student understanding and ... students' ability to solve open-ended, less traditional calculus problems" (Burkam, 1995, p. 1), in the study of Michigan Calculus. In clinical interviews, students in both traditional and reformed classes at UM were asked to solve several calculus problems and to write essays explaining concepts. Students from the new curriculum showed some small, marginally significant advantages over the students from the traditional course. Since one potential outcome of cooperative learning is to facilitate problem-solving skills and higher-order reasoning, it might be that student learning in our study differed between the cooperative and traditional sections in ways that were not measured by the course exams. As for cooperative learning specifically, Owens noted that

> It is here ... that current research and practice involving cooperative learning ... have yet to come fully into "sync" with curricular reform. Simply put, much of the body of cooperative learning research tends to measure achievement in terms that reflect traditional curriculum outcomes—particularly test scores on conventional standardized tests of basic skills. . . . A generation of cooperative learning research and practice that fully incorporates new mathematical goals and values, in a cooperative environment, seems more opportune than expending energy on applying cooperative learning to traditional curricula and assessment practices.
> (Owens, 1995, p. 176)

As new goals are pursued in the teaching of calculus, we need to consider which assessments will best measure progress toward those goals; in many cases, the best measures might differ from our traditional models for testing.

Comparisons between cooperative and traditional instruction have established that cooperative learning is an effective pedagogy, and that research needs to turn to comparisons among cooperative learning methods (Cohen, 1994). Further research needs to uncover which models of cooperative learning can lead to which effects when implemented at the post-secondary level, and for which students. Given the importance in our society of working in groups, even if group work does not improve traditional outcomes, neither does it harm them. We may be helping students develop an important skill by training them to work with others to solve complex problems. As Owens said, " ... there is little point in investigating the question of 'whether or not' cooperative learning should be used. As a model of community and work outside the school . . . cooperative learning is simply a phenomena [sic] whose time has come" (Owens, 1995, p. 174).

Calculus reform involves complex changes, of which cooperative learning is just one component. In evaluating any type of calculus reform, if a mathematics department wants to implement those changes which will be most effective for its population of students, in a cost-effective manner, then they need to know which types of reforms produce which types of effects, and for which students. Having focused for a decade on the question "Does this work?" researchers now need to move on to the next generation of studies, attempting to answer the question "Why and how does this work, and for whom?" Understanding how calculus reform works will require researchers to focus on more specific, detailed questions. For example,

- Which changes lead to which effects? Are all of these changes necessary? Which are sufficient, for which effects?
- What impact do these reforms have on the success and retention of members of under-represented groups?
- What types of technology engage students in calculus in meaningful ways? Is it the technology, or the applied problems themselves, that really matter? What role do calculators play in shaping the visual representations adopted by reformed calculus students?
- Which models of cooperative learning are effective in calculus instruction? Are the improved attitudes found in many reformed courses the result of cooperative learning alone?
- What role do student projects have in shaping students' ideas about mathematics?
- How can we best assess student understanding? What do assessment instruments (exams, lab reports, research papers, etc.) tell us? How do those instruments enhance student understanding?
- For programs that have been successful in specially-selected populations, can they work for other populations? Can they be extended to work for the general student population?
- Can resource-intensive projects be implemented in a sustainable way?

Answering these questions will entail incorporating more diverse research methods. More controlled quantitative (statistical) studies need to be conducted, and interpreted accordingly. Increased use of student interviews and other qualitative research methods could provide richer, more complete descriptions of student learning and attitudes. Mathematics instructors need to foster alliances with educational researchers, in order to take advantage of a wider range of available research tools. Some research in these directions is underway (for example, see the Research in Undergraduate Mathematics Education Community Web Site at `www.cs.gsu.edu/~rumec/`). Further studies that include other theoretical perspectives will give the mathematics community a more complete picture of how the various types of reforms interact with each other to produce the results that we see today.

In many ways, the entire calculus reform movement can be see as a paradigm shift, from concentrating on instructors and what is taught, to focusing on students and what is learned. This has forced the mathematics community to question every aspect of calculus instruction, and great strides have been made toward improving students' experiences. For the reform movement to continue these gains most effectively, research on the reforms has to make a similar shift. The question is no longer whether changes should be made, but which changes will be most effective in achieving specific goals for specific groups of students, in a way that is economically sustainable.

References

Alexander, B., Burda, A., & Hwang, Y. (1995). *Final formative feedback report: Wisconsin Emerging Scholars program, Math 222* (Tech. Rep.). Madison, WI: Univ. of Wisconsin, LEAD Center.

Alvarez, L., Finston, D., Gehrke, M., & Morandi, P. (1993). Calculus instruction at New Mexico State University through weekly themes and cooperative learning. *PRIMUS, 3*, 85–98.

Armstrong, G., Garner, L., & Wynn, J. (1994). Our experience with two reformed calculus programs. *PRIMUS, 4*(4), 301–311.

Boaler, J. (1997). *Experiencing school mathematics: Teaching styles, sex, and setting.* Buckingham: Open University Press.

Bonsangue, M. (1994). An efficacy study of the calculus workshop model. In E. Dubinsky, A. Schoenfeld, & J. Kaput (Eds.), *Research in collegiate mathematics education I* (pp. 117–137). Providence, RI: American Mathematical Society.

Bonsangue, M., & Drew, D. (1995). Increasing minority students' success in calculus. In J. Gainen & E. Willemsen (Eds.), *New directions for teaching and learning* (Vol. 61, pp. 23–33). San Francisco: Jossey-Bass.

Bookman, J., & Blake, L. (1996). Seven years of Project CALC at Duke University: Approaching a steady state? *PRIMUS, 6*, 221–234.

Bookman, J., & Friedman, C. (1994). A comparison of lab based and traditional calculus. In E. Dubinsky, A. Schoenfeld, & J. Kaput (Eds.), *Research in collegiate mathematics education I* (pp. 101–116). Providence, RI: American Mathematical Society.

Brown, M. (1996). Planning and change: The Michigan Calculus project. In A. W. Roberts (Ed.), *Calculus: the dynamics of change* (MAA Notes no. 39, pp. 52–58). Washington, DC: Mathematical Association of America.

Burkam, D. (1994a). *Cognitive benefits of the new curriculum* (Tech. Rep.). Ann Arbor, MI: University of Michigan.

Burkam, D. (1994b). *The Calculus Evaluation Project, interim progress report* (Tech. Rep.). Ann Arbor, MI: University of Michigan.

Burkam, D. (1995). *The calculus evaluation project: Cognitive benefits of the new curriculum* (Tech. Rep.). Ann Arbor: University of Michigan.

Cohen, E. (1994). Restructuring the classroom: Conditions for effective small groups. *Review of Educational Research, 64*, 1–35.

Darken, B., Kuhn, S., & Wynegar, R. (2000). Evaluating calculus reform: A review and a longitudinal study. In E. Dubinsky, A. Schoenfeld, & J. Kaput (Eds.), *Research in collegiate mathematics education IV* (pp. 16–41). Providence, RI: American Mathematical Society.

Davidson, N. (1971). The small-group discovery method as applied in calculus instruction. *American Mathematics Monthly, 78*(7), 789–791.

Davidson, N. (1985). Small-group learning and teaching in mathematics: A selective review of the research. In R. Slavin (Ed.), *Learning to cooperate, cooperating to learn* (pp. 211–230). New York: Plenum Press.

Davidson, N. (1990). *Cooperative learning in mathematics: A handbook for teachers.* Menlo Park, CA: Addison-Wesley.

Douglas, R. (1986). Proposal to hold a conference/workshop to develop alternate curricula and teaching methods for calculus at the college level. In R. G. Douglas (Ed.), *Toward a lean and lively calculus* (MAA Notes no. 24, pp.

6–15). Washington, DC: Mathematical Association of America.

Dubinsky, E. (1992). A learning theory approach to calculus. In Z. A. Karian (Ed.), *Symbolic computation in undergraduate mathematics education* (MAA Notes no. 24, pp. 43–55). Washington DC: Mathematical Association of America.

Dubinsky, E., & Schwingendorf, K. (1997). Constructing calculus concepts: Cooperation in a computer laboratory. In E. Dubinsky, D. Mathews, & B. E. Reynolds (Eds.), *Readings in cooperative learning for undergraduate mathematics* (MAA Notes no. 44, pp. 251–272). Washington DC: Mathematical Association of America.

Fennema, E., & Sherman, J. (1986). *Fennema-Sherman Mathematics Attitudes Scales: Instruments designed to measure attitudes toward the learning of mathematics by females and males.* Madison, WI: Wisconsin Center for Education Research.

Fullilove, R., & Treisman, P. (1990). Mathematics achievement among African-American undergraduates at the University of California, Berkeley: An evaluation of the mathematics workshop program. *Journal of Negro Education, 59*, 463–478.

Ganter, S. (1994). The importance of empirical evaluations of mathematics programs: A case from the calculus reform movement. *Focus on Learning Problems in Mathematics, 16*(2), 1–19.

Green, K. (1989). A profile of undergraduates in the sciences. *American Scientist, 78*, 475–480.

Hagelgans, N., Reynolds, R., Schwingendorf, K., Vidakovic, D., Dubinsky, E., Shahin, M., & Wimbish, G. (1995). *A practical guide to cooperative learning in collegiate mathematics* (MAA Notes no. 37). Washington, DC: Mathematical Association of America.

Hewitt, N., & Seymour, E. (1992). A long, discouraging climb. *ASEE Prism, 1*, 24–28.

Hilbert, S., Maceli, J., Robinson, E., Schwartz, D., & Seltzer, S. (1993). Calculus: An active approach with projects. *PRIMUS, 3*, 71–81.

Hilton, T., & Lee, V. (1988). Student interest and persistence in science: Changes in the educational pipeline in the last decade. *Journal of Higher Education, 59*, 510–526.

Hughes-Hallet, D., & Gleason, A. (1992). *Calculus.* New York: Wiley.

Johnson, D., & Johnson, R. (1994a). *Learning together and alone: Cooperative, competitive, and individualistic learning* (4th ed.). Boston: Allyn and Bacon.

Johnson, D., & Johnson, R. (1994b). An overview of cooperative learning. In J. S. Thousand, R. A. Villa, & A. I. Nevin (Eds.), *Creativity and collaborative learning: A practical guide to empowering students and teachers* (pp. 31–44). Baltimore, MD: Paul H. Brookes.

Johnson, D., Johnson, R., & Holubec, E. (1991). *Cooperation in the classroom.* Edina, MN: Interaction Book Company.

Johnson, D., Johnson, R., & Smith, K. (1991). *Active learning: Cooperation in the college classroom.* Edina, MN: Interaction Book Company.

Kloosterman, P., & Stage, F. (1992). Measuring beliefs about mathematical problem solving. *School Science and Mathematics, 92*(3), 109–115.

Leitzel, J. (1995). ACRE: Assessing calculus reform efforts. *UME Trends, 6*, 12–13.

Levin, J. (1995). Learning outcomes indicated by quantitative data. In *Final evaluation report on the pilot Wisconsin Emerging Scholars Program: 1993-1994* (pp. 34–48). Madison, WI: University of Wisconsin, LEAD Center.

Marco, G., Abdel-fattah, A., & Baron, P. (1992). *Methods used to establish score comparability on the enhanced ACT assessment and the SAT* (Tech. Rep. Nos. 92–93). New York, NY: College Board.

Millar, S., Alexander, B., Lewis, H., & Levin, J. (1995). *Final evaluation report on the pilot Wisconsin Emerging Scholars program: 1993–1994.* Madison, WI: University of Wisconsin, LEAD Center.

Myers, M., & McCaffrey, J. (1994). *The Emerging Scholars Program at UT-Austin program evaluation 1988–1993. Preliminary report.* Austin, TX: Data Center, UT-Austin.

National Advisory Group, Sigma Xi (1989). *An exploration of the nature and quality of undergraduate education in science, mathematics and engineering: Report of the wingspread conference* (Tech. Rep.). Racine, WI: National Advisory Group, Sigma Xi.

National Science Foundation. (1988). *Changing America: The new face of science and engineering. Interim and final reports, the task force on women, minorities, and the handicapped in science and technology.* Washington, DC: National Science Foundation.

Newmann, F., & Thompson, J. (1987). *Effects of cooperative learning on achievement in secondary schools: A summary of research.* Madison, WI: National Center on Effective Secondary Schools.

Ostebee, A., & Zorn, P. (1990). *Viewing calculus from graphical and numerical perspectives.* Washington, DC: Mathematical Association of America.

Owens, J. (1995). Cooperative learning in secondary mathematics: Research and theory. In J. E. Pedersen & A. D. Digby (Eds.), *Secondary schools and cooperative learning: Theories, models and strategies* (pp. 153–184). New York: Garland Publishing, Inc.

Peterson, P., & Fennema, E. (1985). Effective teaching, student engagement in classroom activities, and sex-related differences in learning mathematics. *American Educational Research Journal, 22*(3), 309–335.

Rogers, P. (1990). Thoughts on power and pedagogy. In L. Burton (Ed.), *Gender and mathematics: An international perspective* (pp. 38–45). London: Cassell.

Ross, S. (1996). Visions of calculus. In A. W. Roberts (Ed.), *Calculus: the dynamics of change* (MAA Notes no. 39, pp. 8–15). Washington, DC: Mathematical Association of America.

Schoenfeld, A. (1989). Explorations of students' mathematical beliefs and behavior. *Journal for Research in Mathematics Education, 20*(4), 338–355.

Schoenfeld, A. (1995). A brief biography of calculus reform. *UME Trends, 6*(6), 3–5.

Schwingendorf, K., McCabe, G., & Kuhn, J. (2000). A longitudinal study of the C^4L calculus reform program: Comparisons of C^4L and traditional students. In E. Dubinsky, A. Schoenfeld, & J. Kaput (Eds.), *Research in collegiate mathematics education IV* (pp. 63–76). Providence, RI: American Mathematical Society.

Seymour, E. (1995). Guest comment: Why undergraduates leave the sciences. *American Journal of Physics, 63*, 199–202.

Seymour, E., & Hewitt, N. (1997). *Talking about leaving: Why undergraduates leave the sciences.* Boulder, CO: Westview Press.

Slavin, R. (1995). *Cooperative learning: Theory, research, and practice.* Boston: Allyn and Bacon.

Smith, A., & Moore, L. (1990). Duke University: Project CALC. In T. W. Tucker (Ed.), *Priming the calculus pump: Innovations and resources* (MAA Notes no. 17, pp. 51–74). Washington, DC: Mathematical Association of America.

Solow, A. (1991). Learning by discovery and weekly problems: Two methods of calculus reform. *PRIMUS, 1,* 183–197.

Springer, L., Stanne, M., & Donovan, S. (1997). *Effects of small-group learning on academic achievement among undergraduates in science, mathematics, engineering, and technology: A meta-analysis* (Tech. Rep.). Madison, WI: National Institute for Science Education.

Stroyan, K. (1993). *Calculus using Mathematica.* Boston: Academic Press.

Teles, E. (1992). Calculus reform: What was happening before 1986? *PRIMUS, 2,* 224–234.

Treisman, P. (1985). *A study of the mathematics performance of black students at the University of California, Berkeley.* Unpublished doctoral dissertation, University of California, Berkeley.

Tucker, A. (Ed.). (1995). *Models that work: Case studies in effective undergraduate mathematics programs* (MAA Notes no. 38). Washington, DC: Mathematical Association of America.

Tucker, A., & Leitzel, J. (Eds.). (1995). *Assessing calculus reform efforts: A report to the community* (MAA report no. 6). Washington, DC: Mathematical Association of America.

Tucker, T. (Ed.). (1990). *Priming the calculus pump: Innovations and resources* (MAA Notes no. 17). Washington, DC: Mathematical Association of America.

Vidakovic, D. (1997). Learning the concept of inverse function in a group versus individual environment. In E. Dubinsky, D. Mathews, & B. E. Reynolds (Eds.), *Readings in cooperative learning for undergraduate mathematics* (MAA Notes no. 44, pp. 175–195). Washington, DC: Mathematical Association of America.

Attitude Scales

The first four of these scales were adapted from the Fennema–Sherman Mathematics Attitude Scales (Fennema & Sherman, 1986). The remaining two were constructed for this study. Items are labeled as positive (+) or negative (−).

Confidence in Learning Mathematics

The Confidence in Learning Mathematics Scale is intended to measure confidence in one's ability to learn and perform well on mathematical tasks. The dimension ranges from distinct lack of confidence to definite confidence. The scale is not intended to measure anxiety and/or mental confusion, interest, enjoyment or zest in problem solving (Fennema & Sherman, 1986, p. 4).

Scale items:

+ Generally I feel secure about attempting to learn mathematics.
+ I am sure I could do advanced work in mathematics.
+ I am sure that I sure that I can learn mathematics.
+ I think I could handle more difficult mathematics.
+ I can get good grades in mathematics.
+ I have a lot of self-confidence when it comes to math.
− I'm no good in math.
− I don't think I could do advanced mathematics.
− I'm not the type to do well in math.

- – For some reason even though I study, math seems unusually hard for me.
- – Most subjects I can handle OK, but I have a knack for messing up in math.
- – Math has been my worst subject.

Attitude Toward Success in Mathematics

The Attitude Toward Success in Mathematics Scale is designed to measure the degree to which students anticipate positive or negative consequences as a result of success in mathematics. They evidence this fear by anticipating negative consequences of success as well as by lack of acceptance or responsibility for the success, e.g., "it was just luck" (Fennema & Sherman, 1986, p. 2).

Scale items:

- + I would be happy to be recognized as an excellent student in mathematics.
- + I'd be proud to be the outstanding student in my math class.
- + I'd be happy to get top grades in mathematics.
- + It would be really great to win a prize in mathematics.
- + Being first in a mathematics competition would please me.
- + Being regarded as smart in mathematics would be a great thing.
- – Winning a prize in mathematics would make me uncomfortable because I'd get too much attention.
- – People would think I was some kind of a nerd if I got A's in math.
- – If I had good grades in math, I would try to hide it.
- – If I got the highest grade in math I'd prefer no one knew.
- – It would make people like me less if I were a really good math student.
- – I don't like people to think I'm smart in math.

Mathematics Usefulness

The Mathematics Usefulness Scale is designed to measure students' beliefs about the usefulness of mathematics currently or in relationship to their future education, vocation, or other activities (Fennema & Sherman, 1986, p. 5).

Scale items:

- + I'll need mathematics for my future work.
- + I study mathematics because I know how useful it is.
- + Knowing mathematics will help me earn a living.
- + Mathematics is a worthwhile and necessary subject.
- + I'll need a firm mastery of mathematics for my future work.
- + I will use mathematics in many ways in my life.
- – Mathematics has no relevance to my life.
- – Mathematics will not be important to me in my life's work.
- – I see mathematics as a subject I will rarely use in my daily life after college.
- – Taking mathematics is a waste of time.
- – In terms of my adult life, it is not important for me to do well in mathematics in college.
- – I expect to have little use for mathematics when I get out of college

Effectance Motivation in Mathematics

The Effectance Motivation in Mathematics Scale is intended to measure effectance as applied to mathematics. The dimension ranges from lack of involvement in mathematics to active enjoyment and seeking of challenge. The scale is not intended to measure interest or enjoyment of mathematics (Fennema & Sherman, 1986, p. 5).

Scale items:

- + I like math puzzles.
- + Mathematics is enjoyable and stimulating to me.
- + When a math problem arises that I can't immediately solve, I stick with it until I have the solution.
- + Once I start trying to work on a math puzzle, I find it hard to stop.
- + When a question is left unanswered in math class, I continue to think about it afterward.
- + I am challenged by math problems I can't understand immediately.
- – Figuring out mathematical problems does not appeal to me.

- – The challenge of math problems does not appeal to me.
- – Math puzzles are boring.
- – I don't understand how some people can spend so much time on math and seem to enjoy it.
- – I would rather have someone give me the solution to a difficult math problem than have to work it out for myself.
- – I do as little work in math as possible.

Beliefs About Mathematics

The Beliefs About Mathematics Scale is intended to measure the degree to which students view mathematics as a creative, thoughtful enterprise, as opposed to a static body of algorithmic knowledge.

Scale items:

- \+ In math, you can be creative and discover things by yourself.
- \+ The math I learn in school is thought-provoking.
- \+ There are often several different ways to solve a math problem.
- \+ Time used to investigate why a solution to a math problem works is usually time well spent.
- \+ In addition to getting a right answer in mathematics, it is important to understand why the answer is correct.
- \+ The underlying mathematical ideas are more important that the formulas.
- – Just about everything important about math is already known by mathematicians.
- – Math problems have one and only one right answer.
- – Math is mostly a matter of memorizing formulas and procedures.
- – To solve math problems, you have to know the exact procedure for each problem.
- – Students who understand the math they have studied will be able to solve any assigned problem in five minutes or less.
- – It doesn't really matter if you understand a math problem, as long as you can get the right answer.

Learning With Others

The Learning With Others Scale is intended to measure the degree to which students believe that working with others is helpful in learning mathematics.

Scale items:

- \+ When I can't understand material in calculus class, I like to ask another student in class for help.
- \+ Studying math with others helps me see different ways to solve problems.
- \+ Talking with other students about math problems helps me understand better.
- \+ I prefer to work with other students when doing math assignments or studying for tests.
- \+ I work harder when I work in a group with other students.
- \+ Math is more interesting when I work with other people.
- – When I become confused about something I'm studying in math, I go back and try to figure it out myself.
- – Math is a solitary activity, done by individuals in isolation.
- – When study math with other students, we don't get much done.
- – I learn math best when I study by myself.
- – When I work on math with other students, I usually end up doing more than my share of the work.
- – It's hard to work with other students on math because some students work faster or slower than others.

Department of Learning and Teaching, Graduate School of Education, Rutgers University, New Brunswick, NJ 08901
E-mail address: abbe.herzig@aya.yale.edu

Department of Mathematics and Computer Science, St. Mary's College of Maryland, St. Mary's City, MD 20686
E-mail address: dtkung@smcm.edu

CBMS Issues in Mathematics Education
Volume **12**, 2003

Calculus Reform and Traditional Students' Use of Calculus in an Engineering Mechanics Course

Cheryl Roddick

ABSTRACT. This study investigated traditional calculus and Calculus & *Mathematica* students' conceptual and procedural understanding of calculus and ability to apply that knowledge to engineering mechanics problems. Task-based interviews, during which students solved calculus-related engineering mechanics problems, were conducted with six students who had completed one of the two calculus sequences and were enrolled in an introductory engineering mechanics course. The Calculus & *Mathematica* students demonstrated a preference for the use of conceptual knowledge of calculus, and the traditional students demonstrated a preference for the use of procedural knowledge of calculus. In addition, the Calculus & *Mathematica* group demonstrated a greater tendency to use calculus in their solution attempts. These outcomes may be a result of the different approaches to teaching calculus that were used in their respective calculus sequences.

As part of the calculus-reform movement, reform projects have sought to redefine the teaching and learning of calculus. A common element that has been observed is the de-emphasis of computations in favor of conceptual knowledge and applications. What have been emerging are calculus courses that place an emphasis on understanding, with the goal that students be able to apply that understanding to other situations. This paper presents findings from an investigation of student understanding of calculus, in a course beyond calculus.

Background

Because calculus is a stepping stone to many other courses, success in future calculus-dependent courses may be determined in part by students' experiences in their calculus courses. Although previous researchers have investigated various aspects of calculus-reform efforts (Beckmann, 1988; Crocker, 1991; Estes, 1990; Galindo-Morales, 1994; Hart, 1991; Hawker, 1987; Heid, 1988; Holdener, 1997; Judson, 1989; Lefton & Steinbart, 1995; Meel, 1998; Melin-Conejeros, 1992; Palmiter, 1986; Park & Travers, 1996; Porzio, 1994), few have evaluated students' ability to apply calculus knowledge to other courses (Armstrong, Garner, & Wynn, 1994).

A handful of studies have led the effort to produce evidence of long-term effects of calculus reform (Bookman & Friedman, 1999; Darken, Wynegar, & Kuhn, 2000; Ganter & Jiroutek, 2000; Roddick, 2001; Schwingendorf, McCabe, & Kuhn, 2000). Darken et al. (2000) conducted a longitudinal study comparing the performance of students using the Ostebee and Zorn calculus reform text to those using traditional texts. One focus of the study was to investigate student performance upon completion of two semesters of either reform or traditional calculus. When analyzing subsequent performance (based on grade received in the course) in differential equations and multivariable calculus, no differences were found among the groups. Bookman and Friedman have undertaken an evaluation of Project CALC, one of the calculus reform projects (1999). No significant differences were found between Project CALC students and traditional students in subsequent courses. Schwingendorf et al. (2000) also investigated student performance in subsequent courses and found no significant difference between groups on mathematics courses beyond calculus, and only a slight advantage for reform students on the group of all courses beyond calculus.

The study described in this article investigated achievement in subsequent calculus-dependent courses as well (Roddick, 2001). Results from the analysis of grades showed a significant difference in the introductory differential equations course favoring traditional students, and a significant difference in the first course of the calculus-based physics sequence favoring students from the calculus reform sequence Calculus & *Mathematica*. When grouped according to success in their calculus sequence, the top third of Calculus & *Mathematica* students performed significantly better than the top third of traditional students in the first and second physics courses and an introductory engineering mechanics course.

Results from these studies are promising to calculus reform advocates in that, at the very least, calculus reform students perform no worse than their traditional counterparts on subsequent courses that require use of calculus. Yet anecdotal evidence has given many instructors reason to believe, or to hope, that their calculus reform students are more successful in calculus-dependent courses than an analysis of grades reveals. More research is necessary on the effects of reform students' calculus background when they are enrolled in other courses. More research, as well as a different type of research, is called for in order to begin to understand the differences in the way the two groups understand and use calculus.

In a study that sought to investigate such differences, students from a reform calculus sequence were compared with traditional students on performance on a common final exam for each of the four calculus courses (Ganter & Jiroutek, 2000). This final exam consisted of six to nine questions that were designed to test basic calculus skills using traditional assessment measures. In the first two courses, involving derivatives and integrals, few differences were found between the two groups. Yet in the final two courses, involving sequences, series, and multivariable calculus, the traditional group outperformed the reform group in many areas. However, no significant differences were found between the groups in nine subsequent courses from the following disciplines: chemistry, electrical engineering, and mathematics.

Quantitative comparisons using grades as the criteria for measurement are invaluable in the effort to determine the effectiveness of calculus reform courses. These studies provide us with an overall picture of the ability of the whole group. The work by Ganter and Jiroutek (2000) adds to this picture by providing evidence on specific calculus topics in which differences exist between the two groups. What is also important, to complement the quantitative investigations, are qualitative studies that seek to develop a deeper understanding of a small group of students. Qualitative studies may shed light on other differences in the calculus knowledge of the two groups as well as how each group applies their knowledge of calculus in subsequent courses.

The following study was designed to investigate the qualitative differences in calculus understanding as students continued their education beyond calculus. The focus of the project described here was to explore the way students from one of the calculus reform projects, Calculus & *Mathematica*, and traditional students use calculus in an introductory engineering mechanics course, a course that requires calculus as a prerequisite and incorporates it into the instruction. Investigations were made on 1) students' conceptual and procedural understanding of calculus, and 2) their ability to apply that knowledge to the engineering mechanics course.

Theoretical Framework

One of the underlying goals of the calculus reform movement is that students will understand calculus well enough to be able to apply that knowledge to other courses that use calculus. The theoretical framework for this study proposes a link between active learning with a conceptual focus and students' ability to apply knowledge to other situations (Salomon & Globerson, 1987; Salomon & Perkins, 1989; Winkles, 1986). Salomon and Globerson attribute students' ability to apply knowledge in other situations to the degree of mindful abstraction that occurs during the initial learning. Mindful abstraction is a term uses to describe "mindful deliberate processes that decontextualize the cognitive elements which are candidates for transfer" (Salomon & Perkins, 1989, p. 124). Salomon and Perkins propose that mindful abstraction can be enhanced by active learning with a conceptual focus.

Other research suggests that transfer may be enhanced by an instructional setting in which conceptual learning is stressed over procedural learning (Mayer, 1974; Mayer & Greeno, 1972; Winkles, 1986). In contrast, learning that is based on procedural knowledge is often a result of varied practice and automatization. According to Salomon and Perkins, this type of learning tends to be context bound, and such automatization can create a barrier to conscious reflection (1989). The learner often focuses on superficial features of the problem situation when making a decision about which knowledge to apply (Lesh, Landau, & Hamilton, 1983). Schoenfeld echoes these ideas in his findings on mathematics learning: curricula that emphasize rules and procedures lead to ineffective transfer (Schoenfeld, cited in Vobejda, 1987).

Some researchers also believe that active learning encourages farther transfer than passive reception (DiVesta & Peverly, 1984; Muthukrishna, 1993). It is precisely the *activity* involved in constructivist learning that seems to facilitate the ability to apply existing knowledge to new problem situations. One reason may be that students who are actively involved in their own learning must formulate concepts and decide how to abstract for themselves instead of being given an abstraction by the instructor.

Supporters of this view believe that learning from a constructivist approach and with a conceptual focus can enhance students' abilities to use knowledge in new situations. This view guided the development of the Calculus & *Mathematica* courses (Davis, Porta, & Uhl, 1994).

Method

Setting. This study, which took place at a large Midwestern university, investigated students' understanding of calculus in the context of courses that require calculus as a prerequisite. An introductory engineering mechanics course, a course that requires at least three quarters of calculus and one quarter of physics, was chosen for investigation in this study. The main applications of calculus in the engineering mechanics course involve the concepts of differentiation and integration.

Subjects. The researcher interviewed six students enrolled in an introductory engineering mechanics course. All of the students were male. Three of them had completed the Calculus & *Mathematica* course sequence and the other three had completed the traditional calculus sequence. All of these students had finished the calculus sequence one or two quarters prior to enrolling in the engineering mechanics course. In all but one case participants received As or Bs in the first two quarters of calculus, which covered the calculus topics used in this study. (Tony, one of the Calculus & *Mathematica* students, received a C in the first calculus course, and a B in the second course.)

Calculus & *Mathematica* Background. The four-quarter Calculus & *Mathematica* sequence has been offered at a large Midwestern university since the fall quarter of 1989 as an alternative to the traditional four-quarter calculus sequence. The text (Davis et al., 1994), which is electronic and interactive, utilizes the computer algebra system *Mathematica.* Class is held daily in a computer laboratory and little emphasis is placed on lectures. The main focus in this course is problem solving, which is done on the computer, and includes a variety of real-life word problems. In solving these problems, students are encouraged to discuss their ideas with each other, resulting in a lively, active class atmosphere. Exams, which include paper-and-pencil computation, are given throughout the quarter, and a cumulative final exam is given on paper. Memorization of formulas and practice in hand computation are given considerably less weight than in a traditional calculus course, and more importance is placed on students' writing and clear explanations.

What sets this calculus reform project apart from other calculus courses is not only the immersion in technology, but also the different approach to teaching calculus. Electronic lessons present a conceptual approach to calculus with an emphasis on problem solving. The lack of emphasis on lecture encourages more participation and responsibility on the part of the learner to construct his or her own understanding. As a result, there is a great amount of experimentation and discussion with classmates, along with teacher encouragement of student exploration. In addition, students are required to explain their homework solutions in detail, which encourages deeper understanding.

Traditional Calculus Background. The traditional four-quarter calculus sequence operated under the lecture-recitation format. At the time the participants of this study were enrolled in calculus, the textbook used was *Calculus* (Finney & Thomas, 1991). Although technology was not incorporated into the instruction, students were permitted to use graphing calculators on quizzes and exams. A good portion of the traditional sequence was spent on procedural proficiency, which was reflected in their homework, quizzes, and exams. It should be said that concepts were also developed during instruction in the traditional sequence; yet the time spent on conceptual understanding was less than in the Calculus & *Mathematica* sequence.

Instruments. Task-based instruments, developed and refined from an initial pilot study, were presented to students during four audio-taped interviews that followed a think-aloud protocol. In order to determine students' knowledge of calculus at the beginning of the course, a preliminary instrument was used that consisted of questions similar to final exam questions from the first and second quarters of calculus. Instruments for the final three interviews contained problems that reflected material in the engineering mechanics course. These problems were chosen because they could be solved in more than one way, and in some cases, the use of calculus was optional. A common interview protocol, which focused on eliciting clear explanation of solutions, was followed for all participants. The protocol included a main question for each problem as well as various prompts that were used to follow up on the responses given.

Data Analysis. During the process of analyzing the student interviews, responses were classified according to whether or not calculus was used, and whether the use of calculus involved procedural knowledge, conceptual knowledge, or a combination of both types of knowledge. The definitions of conceptual and procedural knowledge used were those discussed by Hiebert and Lefevre, who classify conceptual knowledge as "knowledge that is rich in relationships" (1986, p. 3), and describe a network of knowledge where all pieces of information are linked to each other in a web-like fashion. Procedural knowledge is defined as a set of symbols and algorithms, where the essential features include actions or transformations that are connected and executed in a linear, or sequential fashion.

Prior to classifying actual student methods, a list of possible responses to the interview questions was generated. Each response was classified according to whether the focus was procedural (through the use of algorithms) or conceptual. Comparisons were then made between the students from the two different calculus backgrounds on the way in which calculus was used and on the students' ability to apply their calculus knowledge to engineering mechanics problems.

In an effort to achieve rater reliability on the classifications of student solution methods, two peers knowledgeable in the areas involved in the study provided their opinions on a subset of the original student solution sheets. With one of the raters, agreement in classification was achieved on 92% of the problems. With the other rater, 85% agreement in classification was achieved. When agreement was not initially achieved, discussions with the raters resulted in a common classification.

Results

Initial Understanding of the Derivative and Integral. Findings from the first interview questions, which covered only differential and integral calculus knowledge and were not engineering mechanics problems, revealed that the preferred approach of the Calculus & *Mathematica* group was conceptual, while the preferred approach of the traditional group was procedural (Roddick, 2001). See Appendix A for interview questions. An important result from that interview is the group of responses to the following questions: what is a derivative? And what

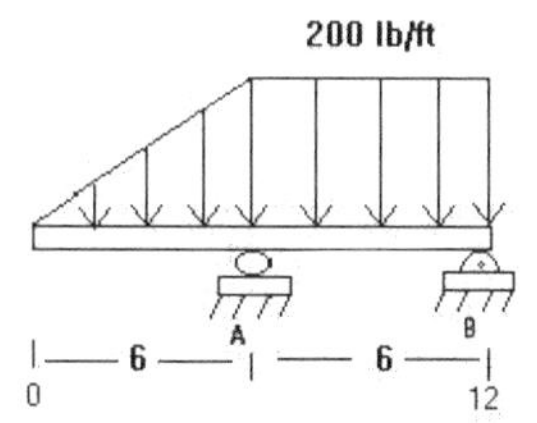

Draw the shear and moment diagrams for the beam. $F_{Ay} = 1400$, $F_{By} = 400$

FIGURE 1

is an integral? These responses are classified in Table 1 in two categories: general definition and specific application. Note that all of the Calculus & *Mathematica* student responses were classified as general definition while the majority of the traditional student responses were classified as specific application. The Calculus & *Mathematica* student responses suggest that the concepts of derivative and integral have been detached from specific examples, and are understood in a more general form. An important question is whether the groups' preferences persist in other courses that require the use of calculus.

Calculus Understanding in Relation to Engineering Mechanics.

The calculus concepts incorporated into the engineering mechanics course are reflected quite nicely in the first two problems. Each of the problems shows a load diagram and students were asked to produce the shear and moment graphs that accompany the load. The knowledge necessary for these problems is that the derivative of the moment M is the shear V, and the derivative of the shear is the negative load w. Engineering mechanics students typically solve these problems in one of three ways: 1) use the ideas of area and slope to construct the shear and moment graphs; 2) find the equation of each part of the load function and integrate twice to find first the shear function and then the moment function; or 3) use the method of sections taught in the class, which involves equations of equilibrium but no calculus. Of the three approaches, the first method is mainly conceptual, the second and third procedural.

Load Problem 1. This problem not only involves ideas of slope and area, but includes an x-intercept of the shear graph, which corresponds to an extreme point on the moment graph. Figure 1 shows the first problem. Students were asked to draw the shear and moment graphs for the beam, given that $F_{Ay} = 1400$, $F_{By} = 400$.

Load problems generally include pins, which secure the object and offer support, and rollers, which only offer support. F_{Ay} and F_{By} represent the vertical forces of the roller and pin, respectively. In Load Problem 1 the pin was on the right and the roller was in the middle, which was not the orientation with which students were familiar. In some cases students flipped the diagram horizontally so that the pin would be at the left. This will be noted during the discussion.

Conceptual solution to Load Problem 1. A possible conceptual solution to this problem is outlined here. (Complete solutions can be found in Appendix B.) The main idea in constructing the shear graph is that the antiderivative of the load is the negative of the shear. Because the graph of the load from $x = 0$ to $x = 6$ is a line with a positive slope, the graph of the shear from $x = 0$ to $x = 6$ would be a concave-down parabola. Furthermore, the change in shear from $x = 0$ to $x = 6$ is

TABLE 1. Responses to Two Interview Questions

Student (Classification)	Response
	What is a derivative?
Traditional	
Gene (App)	It is a tangent of a point of curve. It is the opposite of integral; it takes away a power when produced from the function; the derivative of x^2 is $2x$.
Ed (App)	The derivative tells the fundamental parts that make up an equation....The derivative tells you where it's been, and what its highest and lowest points were.
Nick (Def & App)	At any point on a graph, the derivative of the function will give you the function tangent to it at point (x_0,y_0).
C & M	
Tony (Def)	A derivative measures the rate of change of a function.
Rich (Def)	The derivative is...how the dependent variable changes with respect to the independent variable.
Pat (Def)	A derivative is a measurement of the growth of a function.
	What is an integral?
Traditional	
Gene (App)	It's a tool for solving volume and area; it is the opposite of derivative, it takes the power up when the function is processed; the integral of x^2 is $\frac{x^3}{2}$.
Ed (App)	The integration is the summing of individual equations....If you want to know how acceleration affects velocity, and velocity affects position, you integrate. Integration works your equation to the next level, (i.e.) to a higher order equation.
Nick (App)	It's the summation of an infinite number of divisions between two limits. The integral from x to x_0 of $f(x)dx$ equals the area of the surface underneath the functions and between the two limits. The integral is the antiderivative of an equation.
C & M	
Tony (Def)	The integral can measure the area under a curve. You get it by summing up all the little areas....add up all of the little $f(x)$'s times dx's from here to here (referring to his notation and picture). To get an accurate answer, you want to let these get smaller and smaller and smaller and smaller and add up more and more and more of them.
Rich (Def)	Basically they just summed up all the areas, taking into consideration that this was pretty close....You make the sections smaller so the error is smaller when you look at it.
Pat (Def)	An integral is a measurement of the area under a curve. You make trapezoid boxes and measure the area of this one, add it to this one, and so on. This length is dx. As dx approaches 0, it gets smaller and smaller. Basically, if you make dx really small, the more trapezoids you have, the more accurate your integral should be.

determined by the area under the load from 0 to 6. Because the area is a triangle, it can be determined easily: $\frac{1}{2}(6)(200) = 600$. Thus, the graph of the shear would be a concave-down parabola between (0,0) and(6, -600). At $x = 6$ the vertical force of the roller, F_{Ay}, produces a jump of 1400, up to 800.

The procedure is similar for $x = 6$ to 12. The load is a constant force of 200 lbs./ft. in this interval, so the shear would be a line of slope -200. The change in shear is the area under the load, which is $6(200) = 1200$. So the line starts at (6, 800), crosses the x-axis at $x = 10$, and ends at (12, -400). (This fits with the final force F_{By}, which is 400 upwards.)

For M, it is necessary to graphically find an antiderivative of the shear. From $x = 0$ to 6, M would be a negative cubic, yet the area under the quadratic V is not found as easily as in the load diagram. One way to find the y-value of M at $x = 6$ is to determine the equation of V and integrate. (See the equations in Appendix B.) Another way is to work from right to left. Between $x = 6$ and $x = 12$, V is linear, so M would be quadratic, with the maximum value of M corresponding to the x-intercept of the shear graph. The change in M from $x = 10$ to 12 is the area under V from 10 to 12: $\frac{1}{2}(2)(-400) = -400$. Since there is no moment at the end, M will be 0 at $x = 12$; thus (10, 400) will be the other key point. The change in M from 6 to 10 is $\frac{1}{2}(10-6)(800) = 1600$. Thus the value of M at $x = 6$ is $400 - 1600 = -1200$, which now finishes the graphs.

Calculus & Mathematica Student Responses. All of the Calculus & *Mathematica* students chose to solve this problem by using their knowledge of slope, area, and the shape of the graphs in regard to a function and its antiderivative. Minimal use was made of procedural methods, except when the student considered it necessary.

Pat, one of the Calculus & *Mathematica* students, chose to solve the problem from right to left instead of from left to right. In order to determine the shear diagram, he found the area under the load diagram, separating the load into a triangular and a rectangular portion, and using area formulas for the two basic shapes. Notice that the area under the shear from 0 to 6 is not easy to find because the shear is quadratic. If the problem is approached from left to right, an alternate method must be used to find $M(6)$. Since Pat approached the problem backwards, he was not faced with finding the area under V from 0 to 6. He explained how he arrived at the moment graph:

> The shear is the derivative of the moment. From 6 to 12 the shear has a positive slope, then negative, so I kinda drew it backwards and I knew it ended at 0. Since that's (shear from 10 to 12) negative it had to go down to 0. I found the area of this triangle, which was 400 (shear from 10 to 12), found the area of this (shear from 6 to 10) which was 1600, and subtracted the two and I got -1200 for this lower part and 400 for this crest.

Pat correctly identified the maximum value of M, corresponding to the x-intercept of the shear.

> I looked at this (shear diagram) and I knew the slope was increasing. It was greater increasing because the number was higher here (value of shear at 0) and then it increased but slowlier (sic), and then it (moment diagram) finally peaked here. And then it (x-intercept of shear) started to decrease.

He was looking at the shear graph, whose values represent the slopes of tangent lines of points on the graph of M, since shear is the derivative of the moment. When he referred to the slope as increasing, he really should have said it was the moment function that was increasing, since the shear (slope) values were positive.

Pat's solution approach was largely conceptual and his solution was completed without using a single equation. During the interview, however, he was asked to demonstrate his ability to find equations for V and M, which he was able to do. After finding the equations, he then used his moment equation to verify his value for $M(6)$.

Rich, another Calculus & *Mathematica* student, solved this problem similarly, without the help of equations. Instead, he used his knowledge of area, slope, and the shape of the graphs. Like Pat, he chose to solve this problem from right to left, working with the rectangular portion of the load diagram first. He explained:

> The area under the load function gives how the shear changes.

He used a similar approach to find the moment graph. His solution was correct except for the sign of his graphs; they were the opposite of what they should be. This mistake appeared to occur because he forgot the negative in the relationship between shear and load: the derivative of the shear is the negative of the load.

After his solution, Rich was asked to find equations, which he did, checking both endpoints for each equation he found. (His equations were correct, except for the sign.) Rich correctly related the x-intercept of V to the extreme value of M, and was able to explain the shape of M from $x = 6$ to $x = 12$.

> From 6 to 12 there is a 0 point on the shear at $x = 10$, so it corresponds to a minimum of moment. The slope gradually increases here, that's why it's a concave up parabola, because of the increasing slope.

Note that his sign mistake led him to find a minimum instead of the maximum value and a concave-up parabola instead of a concave-down parabola. Yet he demonstrated an understanding of what the slope of the linear function conveyed about the concavity of the antiderivative (i.e., the quadratic function).

Tony, the third Calculus & *Mathematica* student, used a method similar to those of Pat and Rich in that he used his knowledge of area and slope to construct the graphs. Yet he, unlike the other Calculus & *Mathematica* students, used equations to help him at certain points. Tony described how he found the shear graph:

> The load function is the derivative of the shear and the shear is the derivative of the moment. So if you know the initial point and you know the changes in between, you know the final point.

Tony used his knowledge of area to help him sketch the graph and then proceeded to find equations for V and M by integrating.

For the moment graph, Tony integrated both parts of the shear function. Since he could not find the area under the quadratic function, he needed an equation to find $M(6)$. Recall that Pat and Rich had avoided this problem by working from right to left. Tony's initial moment graph from 6 to 12 did not contain a maximum at $x = 10$ as it should. When I asked if he could check his work another way, he worked to get the moment equation from 6 to 12 by integrating. He said the area under the shear should be the change in moment from 6 to 12. Without taking the positive and negative areas into account he drew a curve without a maximum at $x = 10$. Only when I pointed out the positive and negative areas did he reconsider his initial answer. He then said that the maximum was 400 at $x = 10$, because there was a change of 400 in the shear from 10 to 12.

Traditional student responses. The traditional students' methods of solution were more varied than those of the Calculus & *Mathematica* students. The first traditional student, Nick, used a combination of procedural and conceptual methods. When given this problem, he used the non-calculus method of sections to get equations for V and then integrated to get equations for M. He described how he arrived at his shear graph. (Note that Nick chose to reverse the load function.)

N: I took a cut between 0 and 6 and I found the equation $400 - 200x$. I found the zero at 2.

I: How did you get the shape from 6 to 12?

N: I should have written the equations; I've gone too fast. It would be second order. The load is a negative slope, so when you integrate it's going to be a negative squared. So basically it's going to be an upside down parabola. I was also trying to integrate the load to get the shear. But the only problem is, is that I don't know how to start to write the equation.

For the moment diagram, Tony integrated both parts of the shear function. Since he could not find the area under the quadratic function, he needed an equation to find $M(6)$. Recall that Pat and Rich had avoided this problem by working from right to left. His initial moment diagram from 6 to 12 did not contain a maximum at $x = 10$ as it should. When I asked if he could check his work another way, he worked to get the moment equation from 6 to 12 by integrating. He said the area under the shear should be the change in moment from 6 to 12. Without taking the positive and negative area into account he drew a curve without a maximum at $x = 10$. Only when I pointed out the positive and negative area did he reconsider his initial answer. He then said that the maximum was 400 at $x = 10$, because there was a change of 400 in the shear from 10 to 12.

Traditional student responses. The traditional students' methods of solution were more varied than those of the Calculus & *Mathematica* students. The first traditional student, Nick, used a combination of procedural and conceptual methods. When given this problem, he used the non-calculus method of sections to get the shear equation and then integrated to get the moment equation. He described how he arrived at his shear diagram. (Note that Nick chose to reverse the load function.)

N: I took a cut between 0 and 6 and I found the equation $400 - 200x$. I found the zero at 2.

I: How did you get the shape from 6 to 12?

N: I should have written the equations; I've gone too fast. It would be second order. The load is a negative slope, so when you integrate it's going to be a negative squared. So basically it's going to be an upside down parabola. I was also trying to integrate the load to get the shear. But the only problem is, is that I don't know how to start to write the equation.

His only mistake was that he left out the negative in the relationship between the shear and load. The shear from $x = 6$ to 12 should be a concave up, not concave down, parabola. Since he did not find an equation for V or M for x between 6 and 12, the mistake was not evident to him. When determining the moment graph,

Nick was able to identify the maximum value and relate it to the x-intercept of the shear. He correctly found $M(6)$ as well, from his equation.

His use of conceptual knowledge helped him to determine the shape of the graph and to find where the maximum value of M occurred. However, he was not able to use the area under w or V to determine the change in the shear or moment curve. An equation seemed to be a necessary element in solving this problem. Without an equation, Nick was unable to determine any specific values on the graph; he was only able to give its general shape.

Gene, the second traditional student, also reversed the load. He made several unsuccessful attempts to find the shear graph, due to 1) a misconception that the graph must be symmetric, and 2) trouble finding an equation for V. After a discussion on linear equations, he arrived at the shear equation of $400 - 200x$ for x between 0 and 6. Once he was able to find an equation for V between 0 and 6, he found particular values by substituting in points. He never tried to rethink his symmetry idea, or question why it didn't give him the same line. He finished the shear graph from 6 to 12 without finding another equation.

The moment graph from 0 to 6 was found by integrating to get $400x - 200x^2$. (He made a minor mistake; it should be $400x - 100x^2$.) He noted that the maximum value occurred at $x = 1$, which was correct for his equation, but did not correspond to his x-intercept of its derivative (the shear graph). He made no reference to where the shear crossed the x-axis, which occurred at $x = 2$. Gene was unable to complete the moment graph from 6 to 12 because he could not come up with any equations; at that point he chose to stop.

Gene appeared to be aware of some of the concepts involved, but was not completely comfortable with them. This awareness was evident when he used integration to determine the first half of the shear and moment graphs. Once he found an equation, though, he did not try to connect the maximum value of M with the x-intercept of his shear. His method for finding the shear seemed to be distinct from his method of finding the moment, although the procedures and concepts involved are the same. He, like Nick, was unable to use the area under the curve to sketch the graph or find specific values.

Ed, the third traditional student, used a combination of methods in his solution: two numerical methods, and his understanding of the polynomials produced for each diagram. Ed decided to solve this problem from right to left. His shear graph from $x = 6$ to $x = 12$ was correct except for the sign, and was found by knowing that the shear function would decrease 200 pounds every unit, producing a linear function. Yet the non-constant triangular load caused him confusion in producing the rest of the shear graph. He produced a linear instead of a quadratic curve, unable to reason any differently. This confusion caused him to give up on finding the moment graph from $x = 0$ to 6. He was able, however, to sketch the second part of the moment graph using a numerical adaptation of the method of sections. He explained his thinking:

> The moment is the integration of the shear. If you integrate shear you get moment, because if this line (w) is a constant, this (V) is constant times x. If this (V) is a constant times x, then this (M) is a constant times x^2.

Ed understood which polynomials were generated and how the concept of slope was involved in uniformly distributed loads. However, when the load was not

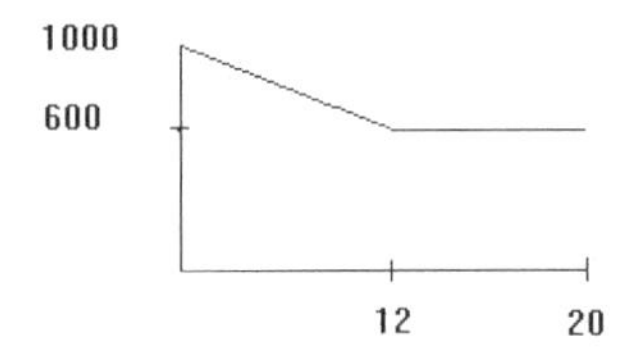

FIGURE 2

uniformly distributed, as in the triangular load in this problem, Ed's conceptual understanding was not versatile enough to be helpful to him. Furthermore, Ed was not able to produce any equations and was unable to use the area under a load or shear curve to determine the change in the shear or moment curve, respectively.

Load Problem 2. Students were asked to solve another problem about shear and moment diagrams, with one interesting change. The load diagram was omitted, and the shear graph was given in its place. Students were provided with the initial moment value, $M(0) = -15,600$, and were asked to sketch the moment graph (see Figure 2). Because the shear function is the derivative of the moment function, students could solve the problem conceptually by first finding the area underneath the shear curve to determine the change in moment, and then using knowledge of the antiderivatives of constant and linear functions to sketch the curve. Students could also solve the problem procedurally by finding equations for V and integrating. Students had not encountered this type of extension previously in their engineering mechanics course. (See solution in Appendix B.)

> [Figure 2] is the shear diagram of a certain beam. Given that $M(0) = -15,600$, sketch the moment diagram.

Calculus & Mathematica student responses. The Calculus & *Mathematica* students approached this problem conceptually and used the area under V to determine the change in M, coupled with an understanding of the function produced by antidifferentiating a constant and linear function. Pat explained his approach:

> The shear is the derivative of the moment so it would start at $-15,600$ and increase to a certain point. The area of this (pointing to the shear diagram) would be what it (M) increases to.

Another student, Rich, explained why M from $x = 0$ to 12 is concave down:

> It's (V) a decreasing value, so the slope's going down. It'll (the slope of the moment function) start out high and decrease.

All Calculus & *Mathematica* students correctly solved the problem, although Pat initially predicted that $M(20) = 0$.

Traditional student responses. Two of the three traditional students, Nick and Gene, found an equation for V and integrated to get an equation for M. At first, Nick drew an incorrect graph without making any computations. When I pointed out his error, he decided to find an equation that corresponded to the shear graph. He found an equation for V between 0 and 12, but not between 12 and 20 because he insisted that the moment always goes to 0 at the end, and he knew it would be a straight line. After discussion, he conceded that the moment may not end at 0 if there was a concentrated moment at the end, but decided there was no way to

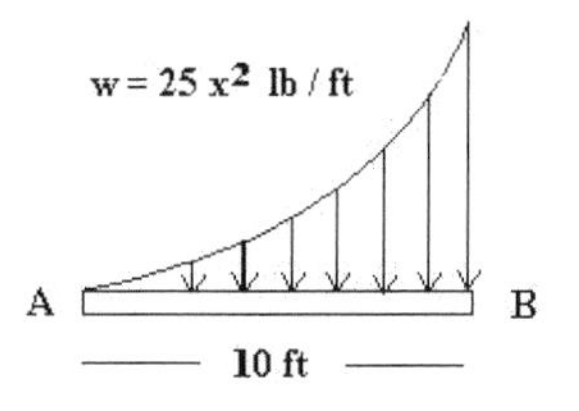

FIGURE 3

tell with the given information. He did not attempt to find the values of $M(12)$ or $M(20)$ or to use an equation for the second part.

Gene also used the procedural method of integrating the shear equation, and he was more successful than Nick. He made a minor mistake integrating, just as he had done in an earlier problem; he forgot to divide by 2 when integrating $\frac{400}{12}x$. The shapes of both parts of the graph were correct, however.

The third traditional student, Ed, was unique in his solution to this problem. His viable approach was to reconstruct the load diagram from the given shear diagram and then apply the method of sections (learned in his engineering mechanics class) to determine key points, but not the equations, of the moment diagram. Ed had an understanding of the shape of the two parts of the moment graph: he explained that the first part from 0 to 12 would be like cx^2 and from 12 to 20 it would be like a constant times x, since the shear is constant in that interval. Unfortunately, even though he had a sound solution approach, he encountered problems with the sign convention in the method of sections as well as deciding on direction and concavity of his parabola, so was not able to produce a correct solution.

Resultant Force Problem. The next problem involved resultant force. Students were asked to determine the magnitude and location of the equivalent resultant force for the force system given. (See Figure 3.)

This problem involves a nonuniformly distributed load for which the magnitude and location of the resultant vector needed to be found. The magnitude can be found by integrating the load function over the given interval. The centroid formula, which involves integration, is used to determine the location of the resultant vector.

All three Calculus & *Mathematica* students and two traditional students exhibited the use of calculus as expected: they used integration to find the magnitude and location of the resultant vector. Ed, the third traditional student, knew he should integrate but could not set up the integral. Tony, a Calculus & *Mathematica* student was the only student who suggested using calculus in any other way.

Tony arrived at a location in his first attempt that he perceived to be incorrect. As a check, he decided to approximate the magnitude and location by summing up small areas. He used his knowledge of the concept of an integral to divide the function into several intervals and to approximate each area by a trapezoid. Although this approach did not provide him with an exact answer, it gave him a good estimate, and when another method gave him an answer close to his estimate, he was able to have confidence in his solution.

This problem contains a nonuniformly distributed load for which the magnitude and location of the resultant vector needed to be found. The magnitude can be

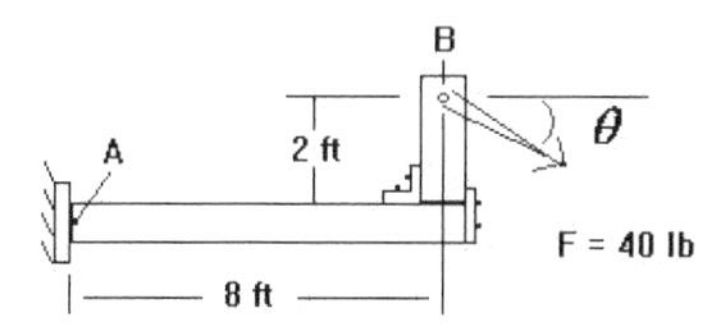

FIGURE 4

found by integrating the load function over the given interval. The centroid formula, which involves integration, is used to determine the location of the resultant vector.

All three Calculus & *Mathematica* students and two traditional students exhibited the use of calculus as expected: they used integration to find the magnitude and location of the resultant vector. Ed, the third traditional student, knew he should integrate but could not set up the integral. Tony, a Calculus & *Mathematica* student was the only student who suggested using calculus in any other way.

Tony arrived at a location in his first attempt that he perceived to be incorrect. As a check, he decided to approximate the magnitude and location by summing up small areas. He used his knowledge of the concept of an integral to divide the function into several intervals and to approximate each area by a trapezoid. Although this approach did not provide him with an exact answer, it gave him a good estimate, and when another method gave him an answer close to his estimate, he was able to have confidence in his solution.

Extrema Problem 1. The following problem requires maximization and minimization of forces and moments. Interview participants were asked to

(1) determine the orientation θ ($0° \leq \theta \leq 180°$) of the 40-lb force **F** so that it produces
  (a) the maximum moment about point A and
  (b) no moment about point A; and
(2) compute the moment in each case. (See Figure 4.)

In a differential calculus course, an extrema problem such as this is typically solved by determining a function and setting the derivative of the function equal to zero. In engineering mechanics, however, students have been taught that the maximum moment about a point is produced when the force is perpendicular to the line connecting points A and B, and no moment is produced when the force is parallel to the line connecting points A and B. This fact comes from the formula for magnitude of the moment: $\|\mathbf{M}\| = \|\boldsymbol{r} \times \boldsymbol{F}\| = rFsin\,\theta$. The maximum occurs when $\sin\theta = 1$; no moment occurs when $\sin\theta = 0$.

Within the engineering mechanics setting, all of the students used the method learned in class. Only one of the students could suggest a way to use calculus in this problem: Tony, one of the Calculus & *Mathematica* students. He was able to incorporate the derivative, its graph, and Taylor approximations to sine and cosine in his solution to this problem. Tony's approach was to find the equation for the moment about point A, then take the derivative to determine the maximum value. His derivative involved sine and cosine, and he was unsure how to solve the equation for θ when it was set equal to zero. He decided a good estimate would result if he used partial Taylor expansions for $\sin\theta$ and $\cos\theta$, and that he could

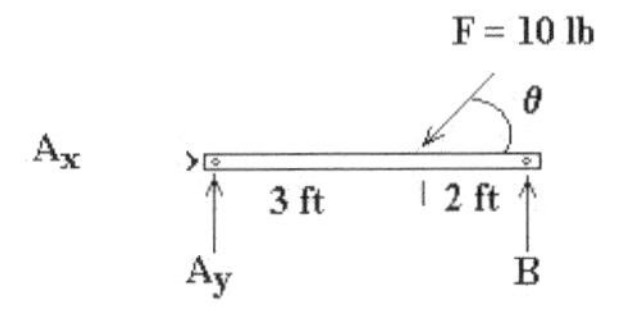

FIGURE 5

get a more exact answer by plotting the derivative on his graphing calculator. His approximated value was close to the answer that he had found without calculus, and close to his solution on the graphing calculator.

Extrema Problem 2. Another extrema problem was posed to the interview participants. (See Figure 5.) Interview participants were asked to 1) find A_x, A_y, and B in terms of θ; and 2) determine for what θ $A = \sqrt{A_x^2 + A_y^2}$ is a maximum or a minimum. This problem can be solved by setting the derivative of $A(\theta)$ equal to zero, graphing the function A, or using knowledge of the sine and cosine functions.

Calculus & Mathematica student responses. Two of the three Calculus & *Mathematica* students chose differentiation as their initial solution method. The other student solved it by inspection and knowledge of the sine and cosine functions. Both of the students had a little trouble finding the derivative by hand, which requires repeated use of the chain rule and/or product rule and is rather messy. Tony, one of the students who used the derivative, also suggested that he could have used his graphing calculator to find the maximum and minimum values of his function. Rich, the Calculus & *Mathematica* student who used his knowledge of the sine and cosine functions, also suggested graphing the function as an alternate method. Although not initially, after some questioning he suggested setting the derivative equal to zero.

Traditional student responses. Each traditional student used a different approach with this problem. Gene solved this problem using reasoning about the position of the forces in the diagram and his knowledge of trigonometric functions. Nick used the numerical guess and check method. Ed was able to think of three ways to solve the problem, one of them involving calculus. He suggested using a graphing calculator, and taking the derivative and setting it equal to zero. When taking the derivative, he could not remember how to differentiate $\cos^2 \theta$. He then remembered two trigonometric identities and finished the problem without the use of calculus.

Discussion

The Relationship between Conceptual and Procedural Knowledge. Student solutions from these engineering mechanics problems provide us with valuable insights about the way students use their knowledge of calculus in other courses. In particular, these problems could be solved in more than one way, and most solutions could have either a conceptual or procedural focus. Although the solutions presented from the six students could be classified as either mostly conceptual or mostly procedural, what seems more important is the way the two types of knowledge were combined to generate a solution. Silver (1986) believes that it is important not to focus on the distinctions between conceptual and procedural knowledge, but rather to focus on the relationship between the two types of knowledge, since problem solving in reasonably complex knowledge domains involves the application of both. Thus, a study of the linkages between the two types of knowledge is advised when investigating problem solving.

The best examples of these linkages came from Load Problem 1. The essence of the problem is to use integration to determine the shear and moment diagrams, yet there are conceptual as well as procedural methods of solution. A mainly conceptual approach involves finding the area under the curve to determine the change in the next diagram over a certain interval. Since only sketches are needed, a student could use his/her knowledge of the antiderivatives of polynomial functions to determine the shape of the curve. Since the given load diagram involves a linear function and a constant function, the area can easily be found. However, the shear diagram involves a quadratic curve, and the area under the curve cannot easily be found. Another method must be used here, and could include finding an equation or working backwards.

Two students, one from each group, implemented both procedures and concepts nicely in their solutions to this problem. Tony, one of the Calculus & *Mathematica* students, began his solution in a conceptual manner, as described in the preceding paragraph, to sketch the shear diagram. Yet, when finding the moment diagram, he could not easily determine the area under the quadratic function, and decided to find the equation of the curve and integrate. His use of procedures was used to complete his conceptual approach.

Nick, a traditional student, began his solution in a procedural manner, finding equations and integrating. Yet he incorporated his conceptual understanding of the shape of the functions produced by antidifferentiating a polynomial function. His conceptual understanding was used to strengthen and verify his procedural solution.

Both of these students chose one main solution approach, either conceptual or procedural, and used the other method to support their primary approach. In fact, most of the students, while demonstrating a preference for either conceptual or procedural solution methods, incorporated both types of knowledge throughout their solution process. Yet, what is quite interesting is that the Calculus & *Mathematica* students in this study demonstrated a preference for the use of conceptual knowledge of calculus, while the traditional students demonstrated a preference for the use of procedural knowledge of calculus (see Table 3). These tendencies may be a result of the different teaching approaches and emphases in their respective calculus sequences.

Use of Calculus in the Engineering Mechanics Problems. Tables 4 and 5 summarize the use of calculus observed in the interview problems. In Table 4, the second column represents procedural use of calculus and in Table 5, the second and fourth columns represent procedural use of calculus; all other columns represent conceptual use. It is evident from the tables that the two groups used calculus in their solutions in a variety of ways.

While all three Calculus & *Mathematica* students were able to demonstrate their ability to solve the problems procedurally, their clear preference was to apply their conceptual knowledge of calculus. The Calculus & *Mathematica* students also were quite successful in solving the load problems, and were able to clearly explain the reasoning behind their use of calculus concepts and procedures.

TABLE 2. Responses to the Load Problems

Student	Response to first load problem	Response to second load problem
Traditional		
Gene	Comb., mostly procedural	Procedural
Ed	Combination	Combination
Nick	Comb., mostly procedural	Procedural
C & M		
Tony	Comb., mostly conceptual	Conceptual
Rich	Combination	Conceptual
Pat	Combination	Conceptual

TABLE 3. Use of Calculus in Load Problems 1 and 2

	Traditional			C & M		
	Gene	Ed	Nick	Tony	Rich	Pat
Integration procedures	1,2		1,2	1	1	1
Knowledge of shape of graph		1,2	1	1	1,2	1
Concept of slope		1		1	1,2	1
Concept of integral as measure of area				1,2	1	1
Relationship of derivative to function maximum			1		1	1

In contrast, the focus of the traditional group was directed more towards the use of actual equations. The traditional students did make use of some conceptual knowledge—the general shape of the graphs and the idea of slope—but used specific equations more than the Calculus & *Mathematica* students. This may be due to the fact that, although traditional students may have known how the concept of slope related to these problems, they did not use the concept of the integral as the area under the curve. Without this way of finding points, an equation or the numerical method of sections was necessary. Surprisingly, two of the three traditional students demonstrated knowledge of the relationship between area and the integral in the preliminary calculus interview (Roddick, 2001). Yet neither of these students used the same idea involving area to help recreate the antiderivative in the engineering mechanics problems.

It is interesting that most did not suggest using the derivative in the first extrema problem, while they did in the second one, another extrema problem. It could be that in the second extrema problem, since it is necessary to first find $A = \sqrt{A_x^2 + A_y^2}$ in terms of θ, it is more natural to use a derivative when they already

TABLE 4. Use of Calculus in Last Three Problems

	Traditional			C & M		
	Gene	Ed	Nick	Tony	Rich	Pat
Integration Procedures	1	1*	1	1	1	1
Approximation of area by trapezoids				1		
Derivative to find max. moment		3*		2,3	3	3
Use of graph to find max. value		3*		2,3	3	
Taylor series to approximate solution				2	3	

* Suggested only

have a function. Also it could be that there is no obvious solution method learned from engineering mechanics to "block" the application of calculus knowledge.

One student stands out when we look at students' use of calculus. Tony, one of the Calculus & *Mathematica* students, was the best able to apply his knowledge of calculus to the engineering mechanics problems. He also happens to be the one who received the lowest grade, a "C," in one of the first two calculus courses. He suggested using both the derivative and the graph of the function to find extreme values, and he knew appropriate applications of the integral. There were also some unexpected uses of calculus. In the integration problem, when his initial integration approach led to an incorrect answer, he decided to approximate the integral by dividing the region into trapezoids and summing the areas of those trapezoids. In the first extrema problem he used partial Taylor expansions for the sine and cosine functions to approximate his solution. These last two uses of calculus are very strong indicators that Tony is knowledgeable enough about calculus to be able to apply it in creative ways.

Conclusions and Future Research

The purpose of this study was to investigate students' understanding and use of calculus in a course beyond calculus. Results demonstrate that the two groups in this study differ on several levels in their use of calculus in an engineering mechanics course.

- The Calculus & *Mathematica* students in this study demonstrated a preference for the use of conceptual knowledge of calculus, while the traditional students demonstrated a preference for the use of procedural knowledge of calculus.
- The Calculus & *Mathematica* students in this study demonstrated stronger ability to discuss the use of calculus involved in the solution of the engineering mechanics problems.

- For problems in which students could choose whether or not to apply calculus, Calculus & *Mathematica* students were more likely than traditional students to incorporate their calculus knowledge.

Although this investigation of six students does not represent an exhaustive evaluation, it does lay a foundation for future investigations in courses beyond calculus. This work has also produced many interesting questions that deserve further investigation.

According to Salomon and Perkins (1989) and Salomon and Globerson (1987) the deliberate process of decontextualizing learning facilitates the ability to use knowledge in a new situation. When students have taken a concept and detached it from the specific contexts in which they have learned it, there is greater likelihood that students will apply that concept to new contexts. With regard to the definitions of derivative and integral given in the preliminary interview (see Table 1), the Calculus & *Mathematica* students in this study have come closer to decontextualizing the concepts of derivative and integral than the traditional students. The three Calculus & *Mathematica* student responses were more general, less bound to specific applications. The question that must be asked is whether average reform calculus students demonstrate tendencies similar to what is reported here. In particular, does the conceptual emphasis of the reform courses translate to conceptual learning in calculus-dependent courses? Does this emphasis translate to increased ability to use calculus in new situations? This study suggests that students from the two groups may demonstrate different strengths in relation to their understanding of calculus. What are the consequences of these different strengths—do the strengths influence the greater success of one group in different types of calculus-dependent courses? And, perhaps most importantly, what can we do in our courses to produce more students like Tony, the Calculus & *Mathematica* student who was able to apply calculus in very unexpected ways?

Another important question is whether the student who completes a reform sequence has a more positive or negative attitude towards mathematics. Does attitude, and confidence in one's ability, affect success in future calculus-dependent courses? Factors such as motivation or learning style preference should be investigated to determine initial differences, as well as affective changes during the four-course sequence, which could impact student attitude towards future calculus-dependent courses.

This study was an effort to begin to address important issues such as these. It is the hope of the author that more studies on reform students' use of calculus in subsequent courses will be conducted in the near future. Such investigations will provide a more complete understanding of the strengths and weaknesses of the calculus reform movement.

References

Armstrong, G., Garner, L., & Wynn, J. (1994). Our experience with two reformed calculus programs. *PRIMUS, 4*(4), 301–311.

Beckmann, C. E. (1988). Effect of computer graphics use on students' understanding of calculus concepts. *Dissertation Abstracts International, 50*(5).

Bookman, J., & Friedman, C. P. (1999). The evaluation of Project CALC at Duke University, 1989-1994. In B. Gold, S. Keith, & W. Marion (Eds.), *Assessment practices in undergraduate mathematics* (MAA Notes no. 49, pp. 253–256). Washington, DC: Mathematical Association of America.

Crocker, D. A. (1991). *A qualitative study of interactions, concept development and problem solving in a calculus class immersed in the computer algebra system mathematica.* Unpublished doctoral dissertation, The Ohio State University.

Darken, B., Wynegar, R., & Kuhn, S. (2000). Evaluating calculus reform: A review and a longitudinal study. In E. Dubinsky, A. H. Schoenfeld, & J. Kaput (Eds.), *Research in collegiate mathematics education* (Vol. IV, pp. 16–41). Providence, RI: American Mathematical Society.

Davis, B., Porta, H., & Uhl, J. (1994). *Calculus & Mathematica.* Reading, MA: Addison Wesley.

DiVesta, F., & Peverly, S. (1984). The effects of encoding variability, processing activity, and rule-example sequence on the transfer of conceptual rules. *Journal of Educational Psychology, 76*(1), 108–119.

Estes, K. (1990). Graphic technologies as instructional tools in applied calculus: Impact on instructor, students, and conceptual and procedural achievement. *Dissertation Abstracts International, 51*, 1147A.

Galindo-Morales, E. (1994). *Visualization in the calculus class: Relationship between cognitive style, gender, and use of technology.* Unpublished doctoral dissertation, The Ohio State University.

Ganter, S., & Jiroutek, M. (2000). The need for evaluation in the calculus reform movement: A comparison of two calculus teaching methods. In E. Dubinsky, A. H. Schoenfeld, & J. Kaput (Eds.), *Research in collegiate mathematics education* (Vol. IV, pp. 42–62). Providence, RI: American Mathematical Society.

Hart, D. (1991). Building concept images: Supercalculators and students' use of multiple representations in calculus. *Dissertation Abstracts International, 52*(12), 4254A.

Hawker, C. M. (1987). The effects of replacing some manual skills with computer algebra manipulations on student performance in business calculus. *Dissertation Abstracts International, 47*, 2934A.

Heid, M. K. (1988). Resequencing skills and concepts in applied calculus using the computer as a tool. *Journal for Research in Mathematics Education, 19*(1), 3–25.

Holdener, J. (1997). Calculus & Mathematica at the united states air force academy: Results of anchored final. *PRIMUS, 7*(1), 62–72.

Judson, P. T. (1989). Effects of modified sequencing of skills and applications in introductory calculus. *Dissertation Abstracts International, 49*, 1397A.

Lefton, L., & Steinbart, E. (1995). Calculus & *Mathematica*: An end-user's point of view. *PRIMUS, 5*(1), 80–96.

Lesh, R., Landau, M., & Hamilton, E. (1983). Conceptual models and applied mathematical problem-solving research. In R. Lesh & M. Landau (Eds.), *Acquisition of mathematics concepts and processes* (pp. 263–343). New York, NY: Academic Press.

Mayer, R. E. (1974). Acquisition processes and resilience under varying testing conditions for structurally different problem-solving procedures. *Journal of Educational Psychology, 66*, 644–656.

Mayer, R. E., & Greeno, J. G. (1972). Structural differences between learning outcomes produced by different instructional methods. *Journal of Educational*

Psychology, *63*(2), 165–173.

Meel, D. (1998). Honors students' calculus understandings: Comparing calculus & mathematica and traditional calculus students. In A. H. Schoenfeld, J. Kaput, & E. Dubinsky (Eds.), *Research in collegiate mathematics education* (Vol. III, pp. 163–215). Providence, RI: American Mathematical Society.

Melin-Conejeros, J. (1992). The effect of using a computer algebra system in a mathematics laboratory on the achievement and attitude of calculus students. *Dissertation Abstracts International*, *53*(7), 2283A.

Muthukrishna, J. (1993). Training mathematical reasoning: Direct explanation versus constructivist learning. *Dissertation Abstracts International*, *53*(11), 3834A.

Palmiter, J. R. (1986). The impact of computer algebra systems on college calculus. *Dissertation Abstracts International*, *47*, 1640A.

Park, K., & Travers, K. (1996). A comparative study of a computer-based and a standard college first-year calculus course. In J. Kaput, A. H. Schoenfeld, & E. Dubinsky (Eds.), *Research in collegiate mathematics education* (Vol. II, pp. 155–176). Providence, RI: American Mathematical Society.

Porzio, D. T. (1994). *The effects of differing technological approaches to calculus on students' use and understanding of multiple representations when solving problems.* Unpublished doctoral dissertation, The Ohio State University.

Roddick, C. (2001). Differences in learning outcomes: Calculus & *Mathematica* vs. traditional calculus. *PRIMUS*, *11*(2), 161–184.

Salomon, G., & Globerson, T. (1987). Skill may not be enough: the role of mindfulness in learning and transfer. *International Journal of Educational Research*, *11*(6), 623–637.

Salomon, G., & Perkins, D. N. (1989). Rocky roads to transfer: Rethinking mechanisms of a neglected phenomenon. *Educational Psychologist*, *24*(2), 113–142.

Schwingendorf, K. E., McCabe, G. P., & Kuhn, J. (2000). A longitudinal study of the C^4L calculus reform program: Comparisons of C^4L and traditional students. In E. Dubinsky, A. H. Schoenfeld, & J. Kaput (Eds.), *Research in collegiate mathematics education* (Vol. IV, pp. 63–76). Providence, RI: American Mathematical Society.

Winkles, J. (1986). Achievement, understanding, and transfer in a learning hierarchy. *American Educational Research Journal*, *23*(2), 275–288.

Appendix A: Interview 1 Questions

1. Find the local maximum and minimum values of $f(x) = x^3 - 3x^2 + 4$.
2. Here are the plots of two functions. One is the derivative of the other. Determine which is which and explain your reasoning.

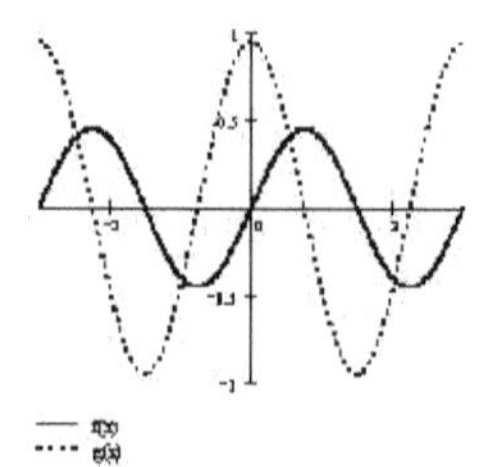

3. Here is the graph of a function $f(x)$ for values of x from -4 to 4. Find $\int_{-4}^{4} f(x)dx$. Explain your work.

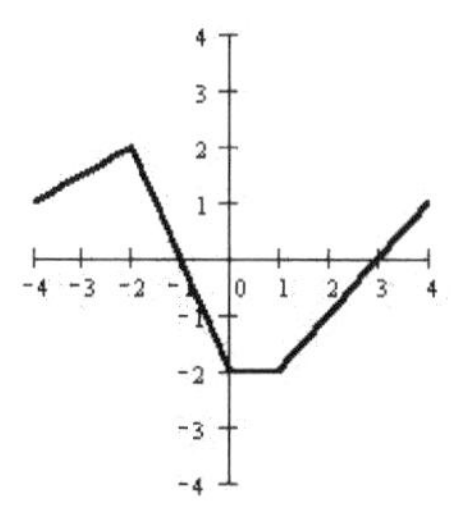

4. Find the area of the region enclosed by the curve $y = \sqrt{2x}$ and the lines $y = 0$, $x = 0$, and $x = 3$.

5a. What is a derivative?

5b. What is an integral?

Appendix B: Solutions

Load Problem 1.

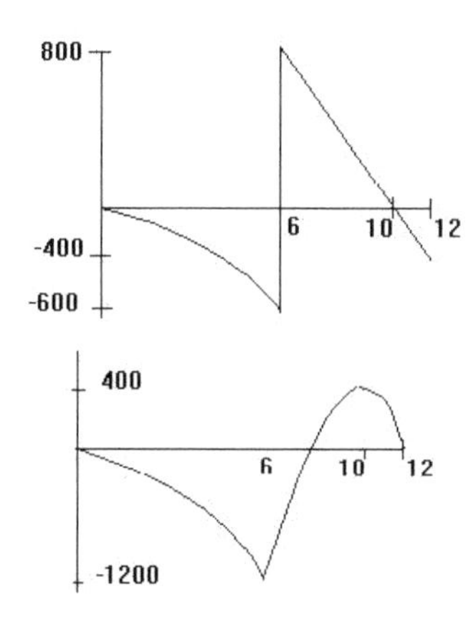

Relationship between M, V, and w.

$$dM/dx = V$$

$$dV/dx = -w$$

Load equations.

$$w(x) = \begin{cases} \frac{200}{6}x & \text{if } 0 \leq x < 6 \\ 200 & \text{if } 6 \leq x \leq 12 \end{cases}$$

Shear equations.

$$V(x) = \begin{cases} \frac{-200}{12}x^2 & \text{if } 0 \leq x < 6 \\ \text{-200(x-6) + 800} & \text{if } 6 \leq x \leq 12 \end{cases}$$

Moment equations.

$$M(x) = \begin{cases} \frac{-200}{36}x^3 & \text{if } 0 \leq x < 6 \\ -100(x-6)^2 + 800(x-6) - 1200 & \text{if } 6 \leq x \leq 12 \end{cases}$$

Load Problem 2.

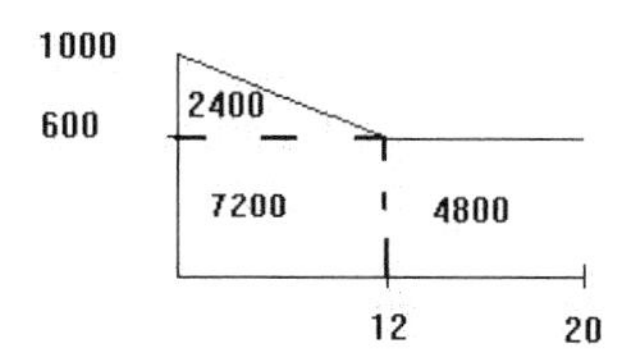

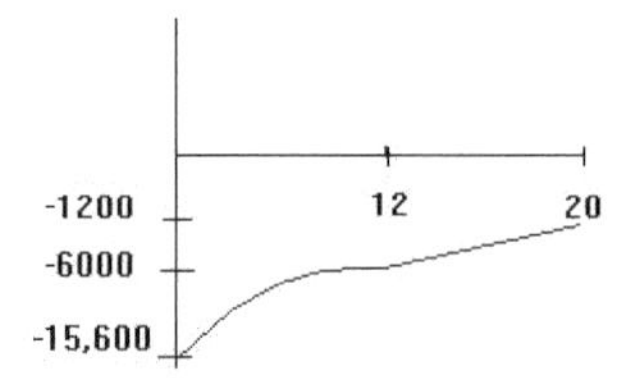

Relationship between M and V.

$$dM/dx = V$$

Shear equations.

$$V(x) = \begin{cases} 1000 - \frac{400}{12}x & \text{if } 0 \leq x < 12 \\ 600 & \text{if } 12 \leq x \leq 20 \end{cases}$$

Moment equations.

$$M(x) = \begin{cases} \frac{-400}{24}x^2 + 1000x - 15,600 & \text{if } 0 \leq x < 12 \\ 600(x-12) - 6000 & \text{if } 12 \leq x \leq 20 \end{cases}$$

Resultant Force Problem. Magnitude: $A = \int_0^{10} 25x^2\,dx = 8333.3$ lbs

Location: $\frac{\int x\,dA}{\int dA} = \frac{\int_0^{10} x25x^2\,dx}{\int_0^{10} 25x^2\,dx} = 7.5$

Extrema Problem 1.
Solution a.

$$M_A = 2(40\cos\theta) + 8(40\sin\theta)$$

$$f(x) = 80\cos\theta + 320\sin\theta$$

$$f'(x) = -80\sin\theta + 320\cos\theta$$

$$80\sin\theta = 320\cos\theta \Rightarrow \theta = 76$$

Solution b.

$$f(x) = 0 \Rightarrow 80\cos\theta + 320\sin\theta = 0 \Rightarrow \theta = 166$$

Extrema Problem 2.

$$\sum F_x = 0 \Rightarrow A_x = 10\cos\theta; \sum F_y = 0 \Rightarrow A_y + B = 10\sin\theta$$

$$\sum M_A = 0 \Rightarrow 3(10\sin\theta) = 5B \Rightarrow B = 6\sin\theta; A_x = 10\cos\theta; A_y = 4\sin\theta$$

$$A(\theta) = \sqrt{16\sin^2\theta + 100\cos^2\theta}$$

Take the derivative and set it equal to 0. Maximum occurs for $\theta = 0°$, minimum occurs for $\theta = 90°$.

Mathematics and Computer Science Department, San Jose State University, San Jose, California 95192

Current address: Mathematics and Computer Science Department, San Jose State University, San Jose, California 95192

E-mail address: roddick@mathcs.sjsu.edu

CBMS Issues in Mathematics Education
Volume **12**, 2003

Primary Intuitions and Instruction: The Case of Actual Infinity

Pessia Tsamir

ABSTRACT. This study examines the tendencies of 121 participants who had taken a Cantorian set theory course and 71 participants who had not, to claim that one-to-one correspondence, inclusion, and "single infinity" are acceptable criteria for comparing the number of elements in two infinite sets. It also investigates the participants' tendency to accept one-to-one correspondence solutions provided to comparison tasks, and the effects of these solutions on participants' acceptance of the three above mentioned criteria. The findings show some significant differences between participants who had taken a set theory course and those who had not. Participants who had taken the course frequently accepted one-to-one correspondence both as a general method for comparing the number of elements in two infinite sets and, when a correspondence was provided, for specific comparison tasks. Nevertheless, a substantial number claimed inclusion and single infinity were also suitable for comparing the sizes of infinite sets.

Introduction

The study described here examined the impact of a Cantorian set theory course[1] on prospective teachers' awareness that one-to-one correspondence is a suitable criterion for comparing infinite sets,[2] i.e., prospective teachers' tendency to accept examples of "set A has a one-to-one correspondence with set B, so A and B have the same cardinality."

More specifically, the study was aimed at investigating the tendencies of participants who had taken a traditional set theory course, and of those who had not, to regard one-to-one correspondence, inclusion, and "single infinity" (i.e., all infinite sets are equivalent because there is only one type of infinity) as suitable criteria for comparing infinite sets, and to accept given comparisons that use one-to-one correspondence.

[1]In Israel, topics related to set theory in elementary, secondary, and teacher education courses are often limited to "elementary set theory" (sets, subsets, union, intersection, etc.). By "Cantorian set theory course," I mean one that includes orders, well ordering, cardinals, ordinals, etc.).

[2]I write "comparing infinite sets" for "comparing the number of elements in two or more infinite sets."

Clearly, comparison of infinite sets is connected to the facet of infinity which "we meet . . . when we regard the totality of numbers 1, 2, 3, 4, . . . itself as a complete unity, or when we regard the points of an interval as a totality of things which exists at once. This kind of infinity is known as *actual infinity*" (Hilbert, 1924/1989). Still, acceptance of the notion of actual infinity by the mathematical community was not smooth and simple. When Cantor first presented his set theory in the late 19th century, introducing infinite sets and infinite numbers, he met with a mixture of hostility and enthusiasm. Although a number of great mathematicians and philosophers, such as Poincaré and Kronecker, fiercely objected to it, others, like Hahn, Russell, Hilbert, and Dedekind enthusiastically accepted it. And those who accepted Cantor's notion of actual infinity, including Cantor himself, felt the need to stress that these notions contradict everyday experience. Russell, for instance, noted the natural tendency to reject the basic attribute of infinite sets—that they have at least one proper subset of the same cardinality. He mentioned that one reason for rejecting the notion of actual infinity would be our tendency to accept that "the whole cannot be similar to one of its parts" as doubtless and self-evident (Russell, 1903/1950, p. 360). Cantor further asserted that all the arguments rejecting infinite numbers were either rooted in reasons relevant only to finite ones, or in the failure to understand the need for special, differentiating characteristics for the new, infinite entities (Cantor, 1885/1950, p. 74).

Cantor's awareness of the psychological difficulties that hinder the acceptance of actual infinity was probably the result not only of academic analysis, but also of his own experience. Evidence of Cantor's hesitation and the conflict he experienced between the finite constraints and the logical implications of his analyses can be found in a series of letters he wrote to Dedekind. Fischbein (1987) examined Cantor's inner struggle and explained his dilemma regarding the equivalency between the number of points on a line and on a plane:

> It seems that, while most of the mathematicians of his time were trapped in one intuition, Cantor was trapped in *two* contradictory intuitions: the old, "natural" intuition, according to which two continuous sets of points having a different number of dimensions cannot be equivalent and the new, Cantorian intuition claiming the equivalence of the two sets. (1987, p. 26)

Fischbein used the term intuition to mean a special type of cognition characterized by self-evidence, immediacy, stability and coerciveness. He viewed intuitions as implicit extrapolations which are indirectly and tacitly made from a limited amount of information, and whose acceptance is accompanied by a powerful feeling of certainty. Intuitions offer a unitary global view, which expresses a general property that is perceived through a particular experience (Fischbein, 1987, 1993).

Fischbein distinguished between two types of intuitions: Primary intuitions, which "develop in individuals independently of any systematic instruction, as an effect of their personal experience" (Fischbein, 1987, p. 202), and secondary intuitions, which are new logically-based interpretations, which develop only under special systematic instructional intervention, and which compete with primary intuitions.

Fischbein also defined some mathematical and scientific notions or representations that trigger scientifically incorrect ideas as non-intuitive or counter-intuitive concepts (1987, p. 10). He regarded the Cantorian concept of actual infinity as just such a notion. Indeed, research in mathematics education over the last two decades has indicated that, when comparing infinite sets, students intuitively use a wide range of criteria for their comparisons. This research also reported that one-to-one correspondence, the criterion used in Cantorian set theory, is generally neglected (Borasi, 1985; Moreno & Waldegg, 1991; Fischbein, Tirosh, & Melamed, 1981). Moreover, students often reach contradictory conclusions by using various criteria, such as "single infinity," "inclusion," and "incomparability." By following these criteria, students conclude simultaneously, "All infinite sets have the same number of elements," yet "A set that is included in another set has fewer elements than that set," and occasionally also, "It is impossible to compare infinite sets." Students usually remain, however, unaware of these contradictions (Tsamir & Tirosh, 1992, 1994). Another major finding was that students' responses are representation-dependent and are influenced by irrelevant, visual aspects (Duval, 1983; Tirosh & Tsamir, 1996; Tsamir & Tirosh, 1994).

There are several striking similarities between young children's first attempts to understand how many objects there are in a finite collection and students' attempts to understand the cardinality of infinite sets. In his seminal work on the child's conception of number Piaget (1941/1969) examined children's ability to identify one-to-one correspondence in pairs of given finite sets, and the impact of the perceptual, dimensional appearance of the elements on children's initial grasp of equivalency. Children were asked, for instance, to put beads in a container, one by one, at the same time as the experimenter was putting beads into another container. They were then twice asked whether the total quantities of beads in the two containers were the same, once when the shapes of the containers were identical and subsequently when they were not. All children correctly argued that the numbers of beads in the identical containers were the same, some children explained that two corresponding beads were put in at the same time, and others explained the equality in terms of the identical shape of the containers. However, when asked about the beads in the differently shaped containers, all children claimed that the numbers of beads were not equal. This conclusion appeared to be based on the perceptual differences between the two containers, because the children mentioned the different levels of the beads in the two containers, claiming that "there'll be more here . . . because it's higher" (p. 30). Piaget concluded that in this case, "perception completely over-rules the correspondence" (p. 29). Moreover, at this stage, the young children were unaware of the contradiction between their solutions. No conflict arose in their minds between their negating "equal number" answers, triggered by the correspondence, and "unequal number of elements" answers elicited by the perceptual appearances.

As stated before, it seems that the difficulties identified by Piaget (1941/1969) in children's early grasp of the cardinality of finite sets do, in several aspects, resemble students' difficulties in grasping the cardinality of infinite sets. In both cases, participants' decisions about the numbers of elements in the sets are sensitive to irrelevant, dimensional aspects and they are unaware of the contradictions between their own responses. The difficulties and contradictions related to children's grasp of the cardinality of finite sets are resolved with age. So the question that naturally arises is: Is this the case for infinite sets as well? Are difficulties and contradictions related to students' grasp of the cardinality of infinite sets also resolved with age?

The common view of those leading 19th and early 20th century mathematicians who accepted the notion of actual infinity was that the learners' problems, rooted in the counter-intuitive nature of infinity, could easily be overcome. They claimed that only a little practice would be needed to gain "true and better instincts" of the infinite. The presentation of the axioms, definitions, theorems, etc., should be accompanied by a simple declaration that infinity has no existence in reality (Russell, 1916/1980, p. 1560). Fraenkel (1953, p. 13) for instance, stated, "difficulties would evaporate the moment it is clarified that there is no such case in reality." One may wonder—How much is "a little practice"? How are "true and better instincts" acquired? How quickly do the difficulties "evaporate"?

One way to answer the latter questions is by surveying the research that examined and evaluated different relevant interventions. Mathematics education researchers, when trying to formulate appropriate interventions and examine their influence on students' performance in Cantorian set theory, reported a number of attempts to promote students' acceptance of one-to-one correspondence as *the* criterion that underlies any method for comparing infinite sets. For instance, Sierpinska and Viwegier (1989) investigated how the intervention most commonly used by mathematics teachers, presenting the correct solution, would affect the acceptance of the one-to-one correspondence criterion. They presented two students (10 and 14 years old) with a correct procedure to create a one-to-one correspondence between points in pairs of different segments. While the younger student tended to accept the "pairing idea," the other tended to reject it, finding it counter-intuitive.

Other mathematics educators investigated the impact of interventions on students' acceptance of the one-to-one correspondence criterion, using activities that were based on research findings regarding intuitions about infinity. Tirosh (1991) presented one such intervention as successful in promoting students' understanding of the notion of infinity. She described a set theory course for secondary school students, which used, among other things, the cognitive conflict approach followed by a whole-class discussion of students' intuitive ideas. Another interesting intervention centered on different representations of mathematically identical infinite sets. For example, two possible representations of the set of natural numbers and that of perfect squares are:

$$A = \{1, 2, 3, 4, 5, \dots\} \quad B = \{1, 4, 9, 16, 25, \dots\}$$

and

$$A = \{1, 2, 3, 4, 5, \dots\} \quad C = \{1, 2^2, 3^2, 4^2, 5^2, \dots\}$$

The aim of this intervention was to spark cognitive conflict and thus promote high school students' awareness of the inconsistency of their ideas when they compare infinite sets (Tsamir & Tirosh, 1999). The findings indicated that a non-negligible number of students gave different answers when comparing two sets, depending on the different representations, but when students grasped that the pairs of representations corresponded to the same sets, they suggested different ways to resolve the "contradiction." Some chose a single "preferred criterion" as the only criterion for comparing infinite sets, others tried to explain why this incompatibility was "harmless," and a number of students claimed that they needed to study the relevant theorems.

Another way to probe possible interventions and examine their impact on students' tendency to grasp one-to-one correspondence as the only suitable criterion of the three presented for comparing infinite sets, is by investigating how mathematics teachers actually teach Cantorian set theory courses. Nowadays, in Israel, prospective secondary mathematics teachers study the concept of actual infinity at teachers college. This concept is usually introduced in the Cantorian set theory course at the time when the Zermelo-Fraenkel axioms are discussed. It is taught through traditional lecturing in the form of a consistent and sequential presentation of axioms, basic notions, definitions, and theorems. In this spirit, comparison tasks are discussed, and one-to-one correspondence and the examination of the cardinalities of the sets are presented as methods for determining whether or not two infinite sets have the same number of elements. Nevertheless, students' primary intuitions are usually not taken into account either in planning instruction or in teaching. Students' intuitive tendency either to compare infinite sets by inclusion or to consider infinity to be a single entity is usually neglected.

Examination of the various suggested and applied interventions reveals that the participants in all the above mentioned research studies were younger than the students who study Cantorian set theory as part of their curriculum, and that they had not taken such a course. There is no documented research regarding tertiary students' grasp of actual infinity, either before or after having studied Cantorian set theory. Thus, it seemed important to investigate impact of age and study of Cantorian set theory on students' comparisons of infinite sets.

This study, then, focused on teachers college students, some of whom had already studied the required course in Cantorian set theory and some of whom had not. The study examined their respective tendencies to accept the criteria "one-to-one correspondence," "inclusion," and "single infinity" for the comparison of infinite sets. The research questions were: (1) Do participants regard one-to-one correspondence, inclusion or single infinity as criteria suitable for the comparison of infinite sets? (2) Do prospective teachers accept illustrated ways of applying one-to-one correspondence for the comparison of infinite sets? (3) Do the illustrated solutions affect participants' acceptance of one-to-one correspondence, inclusion, and single infinity as criteria for the comparison of infinite sets? and (4) Will there be significant differences between participants who already studied the topic and those who have not?

Method

Participants. Participants were 181 prospective secondary school mathematics teachers, sampled from Israeli state teachers colleges. Seventy-one of them (denoted N-ST) had not studied Cantorian set theory and 110 (denoted ST) had completed a yearlong Cantorian set theory course three months prior to participating in this study. The course consisted of 24 90-minute sessions. Its syllabus included the following topics: the notion of set, finite and infinite sets, axiomatic development of set theory, relations and operations between sets, functions, the comparisons of the number of elements in infinite sets, cardinal numbers, well-ordered sets, ordinal numbers, axiom of choice, Zorn's lemma, and paradoxes in set theory. This course was taught in a traditional manner with no reference to the participants' intuitions.

Questionnaire. During a mathematics lesson, the participants were asked to fill out a questionnaire,[3] which consisted of the following two parts.

Part I: Acceptability of criteria for comparing infinite sets. Explanations illustrating the notions of one-to-one correspondence, single infinity and inclusion were provided. Subjects were asked to determine whether each of the criteria was suitable for comparing infinite sets.

Sample Problem

> *In class, students suggested different ways for comparing the number of elements in infinite sets.*
>
> *Danny: When set A is included in set B, i.e., set B consists of all the elements of set A and at least one additional element, then the number of elements in set B is larger than the number of elements in set A (the criterion of "inclusion").*
>
> In your opinion, is the criterion of inclusion suitable for comparing the number of elements in infinite sets? Yes / no
>
> Explain your answer.

The criteria single infinity and one-to-one correspondence were presented in a similar way, and participants were asked the same question in each case.

Part II: Acceptability of one-to-one correspondence through illustrated solutions. Four illustrations of the use of one-to-one correspondence to compare given pairs of infinite sets were provided. The solutions were presented in the form of a suggestion made by a student (John) who had understood that one-to-one correspondence was the criterion to be used when comparing infinite sets. For all the cases, participants were asked to state whether they found the suggested solution acceptable and to justify their judgments.

Sample Problem

> *John claimed that one should only use "one-to-one correspondence" as the criterion for the comparison of infinite sets. He provided several examples to illustrate such comparisons.*
>
> *Problem 1*
>
> *When given:* $A = \{1, 2, 3, 4, 5, 6, \dots\}$ $B = \{1/2, 1, 1\,1/2, 2, 2\,1/2, 3, \dots\}$
>
> *The number of elements in set B is equal to the number of elements in set A. Every element in set B is the result of multiplying one element of set A by $1/2$. That is, if x is an element in set A then $1/2x$ is the matching element in set B. No other type of element exists either in set A or in set B.*

[3] Care was taken not to discuss in class the specific problems that were included in the questionnaire.

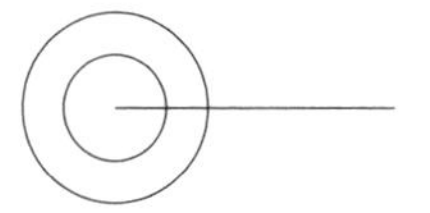

FIGURE 1

In your opinion, is John's solution acceptable? Yes / no

Explain your answer.

The other comparisons were presented in a similar way. All pairs of sets were presented vertically (i.e., one on top of the other). The illustrated comparisons related to five, given pairs of infinite sets. Three examples presented the following pairs of denumerable sets:

1. $A = \{1/2, 1, 1\,1/2, 2, 2\,1/2, 3, \dots\}$
   $B = \{1, 2, 3, 4, 5, 6, \dots\}$
2. $B = \{1, 2, 3, 4, 5, 6, \dots\}$
   $E = \{3, 4, 5, 6, 7, 8, \dots\}$
3. $I = \{4, 8, 12, 16, 20, \dots\}$
   $J = \{1, 4, 9, 16, 25, \dots\}$

One example dealt with a pair of non-denumerable sets of the same cardinality, comparing the number of points on two concentric circles.

Problem 4

When given two concentric circles:

I=*{the points on the small circle}*
U=*{the points on the large circle}*

The number of elements in set I is equal to the number of elements in set U. The two circles consist of an equal number of points. Each ray that originates in point M (the center of both circles) intersects the small circle in exactly one point (one element of set I), and also intersects the large circle in exactly one point (one element of set U), as seen in Figure 1.
That is, if x is an element in set I then y is the matching element in set U. No other type of element exists either in set I or in set U.

Process

The study consisted of four stages.

(1) Stage I: All participants responded to Part I of questionnaire.
(2) Stage II: All participants responded to Part II of questionnaire.
(3) Stage III: Part 1 of questionnaire was re-administered to all participants.
(4) Stage IV: 20 participants were interviewed.

In the first three stages, participants were asked to answer the questionnaire in writing. Stage III was meant to examine whether and how, after having been presented in Part II with examples of correct solutions, these solutions would influence

TABLE 1. Frequencies (in %) of accepting criteria

Criterion accepted	One-to-one correspondence		Inclusion		Single infinity	
Study of set theory	N-ST	C-ST	N-ST	C-ST	N-ST	C-ST
	$n = 71$	$n = 110$	$n = 71$	$n = 110$	$n = 71$	$n = 110$
Stage I	54.3	76.4	63.4	31.8	31.4	17.3
Stage III	73.9	91.6	60.9	32.4	54.3	25.5

the participants' reaction to the questions presented in Part I. The participants were given about 45 minutes for all three stages. They received each subsequent part of the questionnaire only after handing in the previous one.

In the fourth stage, after completing the written assignment, 10 N-ST and 10 C-ST participants, who had all provided insufficient justifications, were interviewed individually. They were asked to further explain their written responses, in order to offer a better insight into their ideas. No additional problems were posed during interviews. A typical interview lasted about 30 minutes. The findings of the interviews are integrated, where suitable, into the Results section.

Results

The results are presented according to the first three stages of the research process.

Stage I. Declaring One-to-One Correspondence, Inclusion, and Single Infinity as Acceptable Criteria. Table 1 shows that in Stage I, one-to-one correspondence was the criterion most frequently accepted by C-ST participants and inclusion was the criterion most frequently accepted by N-ST students. Single infinity was the least accepted criterion by both N-ST and C-ST participants. It should be noted that even among C-ST participants, about 17% still viewed single infinity as suitable for comparing infinite sets and about 32% of them still viewed inclusion as suitable.

This criterion is the best. This type of response emphasized that the suggested criterion must be used for comparing infinite sets. About half of those who accepted one-to-one correspondence presented such a justification. Interestingly, about 20% of the C-ST participants who regarded one-to-one correspondence as a suitable criterion, justified this claim in terms of "power," by claiming that "actually, in order to compare the number of elements one should examine the powers of the sets, but one-to-one correspondence is also quite OK."

Moreover, about 60% of the N-ST and half of the C-ST participants who accepted inclusion expressed the idea that "the use of inclusion is ideal because it allows you to reach definite conclusions." About 40% of the N-ST and half of the C-ST participants, who claimed that single infinity was a suitable criterion, explained this by stating that "infinity is always the same infinity, so obviously you should follow only this criterion."

Practical considerations. These responses expressed a general attitude that any criterion may be applied when suitable. About 30% of the N-ST participants presented this explanation for each of the three criteria. Among the C-ST participants, about 20% of those who accepted one-to-one correspondence, 30% of those who accepted single infinity and 40% of those who accepted inclusion presented this line of

TABLE 2

Problem	1	2	3	4
Correspondence	$n \leftrightarrow 1/2n$	$n \leftrightarrow n+2$	$4n \leftrightarrow n^2$	Rays from circles' center
Chi-square results	$\chi^2 = 22.6$	$\chi^2 = 36.04$	$\chi^2 = 13.2$	$\chi^2 = 20.9$
	DF = 2	DF = 2	DF = 2	DF = 2
Significance	$p < 0.01$	$p < 0.01$	$p < 0.01$	$p < 0.01$

reasoning. Explanations were, for instance, that "not in all cases is there inclusion, but when one set is included in the other, it is only natural to use this criterion" or "it is hard to find matching elements, but when possible, this method should be used to compare the sets."

Participants justified their rejection of the various criteria with four types of explanations:

Direct counter-arguments. These responses rejected a criterion due to its attributes or the nature of infinity. Most frequent claims were: (a) "A criterion that always gives the same conclusion is useless"; this was given by participants who rejected single infinity, about 40% of the C-ST and about 70% of the N-ST; and (b) "Even when I find one-to-one correspondence between matching elements, I can only control a finite number of 'pairs,' so I can never be sure that the matching rule keeps working infinitely." This was presented by about 60% of all participants who rejected one-to-one correspondence.

Indicating another criterion as more suitable for comparing infinite sets. This line of justification was used by 15% of the N-ST participants who rejected single infinity because "one should use inclusion"; and by about 20% of the N-ST participants who rejected inclusion because "there is one, single infinity." Among the C-ST participants, about 25% of those who rejected inclusion and 40% of those who rejected single infinity argued that power is the only correct criterion. Surprisingly, the idea of power even served about 20% of the C-ST participants who rejected one-to-one correspondence. They claimed, for instance, that "We have learned that only the powers of sets indicate the numbers of elements they have. Obviously, in order to compare the number of elements in two sets, one needs to know these numbers."

Practical considerations. These responses excludedng the use of a certain criterion due to its practical limitations. About 30% of all participants justified the rejection of inclusion by claiming that "the criterion is inadequate, as frequently there is no inclusion relationship between the sets." Similarly, about 30% of the N-ST participants who rejected one-to-one correspondence said, "In many cases it is extremely difficult to find a way to match the elements."

Consistency considerations. These responses expressed the need to preserve consistency. Only when rejecting inclusion did a few N-ST and C-ST participants claim that "inclusion is not valid as it can contradict solutions arrived at by using one-to-one correspondence."

Stage II. Acceptance of Illustrations Using One-to-One Correspondence. Figure 2 shows that for each of the four problems that presented sets of the same cardinality, the tendency of C-ST participants to accept the illustrated one-to-one correspondence (about 90%) was significantly higher than that of the N-ST participants (about 60%), see Table 2 and Figure 2.

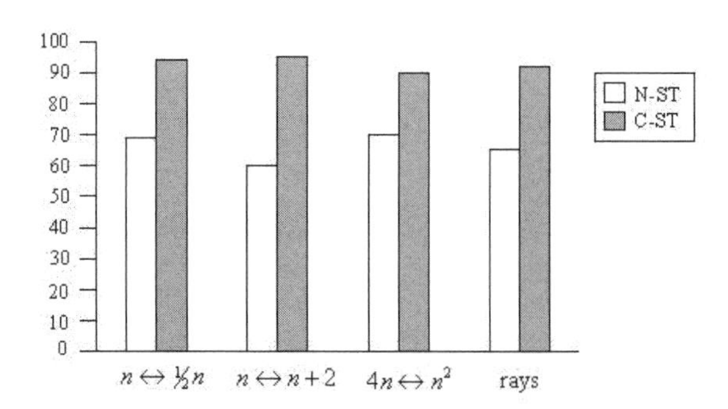

FIGURE 2. Frequencies (in %) of the acceptance of illustrated "one-to-one correspondence" solutions

Most participants did not justify their acceptance or rejection of the given solutions. Nevertheless, among those who did, acceptance of the one-to-one correspondence presented, was usually justified by claiming that:

The suggested idea seems good. The given solution was regarded as being a good, correct way for conducting the given comparison (N-ST participants).

The given solution reminded them of the course studied. The given solution reminded most C-ST participants of the ways to compare infinite sets, which they had studied in the set theory course.

Rejection of the suggested solution, when justified, was usually accompanied by the following lines of reasoning:

Indicating another criterion as more suitable for comparing infinite sets. N-ST participants most frequently explained rejecting the given solutions to the equivalent sets by specifying either inclusion or single infinity as being a preferable criterion. Most prevalently they related to inclusion "One set is included in the other, thus they don't have the same number of elements," but they also mentioned single infinity "one-to-one correspondence is unnecessary. It is known that infinities are always the same."

C-ST participants who rejected the suggested solution, however, usually accepted the equivalency, but said that power and not one-to-one correspondence was the criterion to be used.

Infinite sets are incomparable. This explanation was given for all cases by just a few participants.

Stage III: Re-Declaring One-to-One Correspondence, Inclusion and Single Infinity as Acceptable Criteria (Part I). Table 1 shows that when advancing from responding to Part I in Stage I to responding to the same Part in Stage III there was an increase in the percentage of N-ST and C-ST participants who claimed one-to-one correspondence to be suitable for comparing infinite sets. However, the growing acceptance of one-to-one correspondence was not accompanied by an increased tendency to reject the other criteria. In Stages I and III, similar rates of N-ST participants viewed inclusion as acceptable (about 60%), and the same was true for C-ST participants (about 30%). Surprisingly, more N-ST and more C-ST accepted the single infinity criterion at Stage III than at Stage I.

Usually participants either did not explain or wrote that they had already explained their judgments in Stage I. Participants who justified their claims often referred to the way Part II influenced their reasoning:

Stage II (Part II of the questionnaire) illustrations triggered acceptance of one-to-one correspondence. "I was convinced by the illustrations in Stage II that one-to-one correspondence can be very useful even in cases where I, personally, couldn't provide the matching rule" or "The previous examples [presented in Stage II] reminded me" (used by participants who rejected one-to-one correspondence in Stage I and accepted it in Stage III).

Stage II illustrations triggered rejection of one-to-one correspondence. "All these questions confuse me and make me believe that infinity is really strange. There is probably a single infinity" or "Inclusion seems more reasonable than creating a correspondence between an infinite number of pairs" (used by participants who accepted one-to-one correspondence in Stage I but rejected it in Stage III).

Stage II illustrations had no influence on the grasp of one-to-one correspondence. "From the beginning I viewed one-to-one correspondence as a suitable criterion" (those who consistently accepted one-to-one correspondence in Stages I and III). "Infinity is always the same infinity and that's what I have written in all questions" or "I already answered in all previous questions that infinite sets are incomparable" (a few participants who consistently expressed in both Stages I and III the notion of single infinity to reject one-to-one correspondence, or some others who argued at both parts that "infinite sets are incomparable").

Discussion

The findings of this research indicate that a non-negligible number of participants intuitively regarded the criterion of one-to-one correspondence as suitable for the comparison of infinite sets. When asked in Stage I of the questionnaire to determine acceptability of each of the three criteria, about half of the N-ST and three quarters of the C-ST participants claimed that one-to-one correspondence was suitable for comparing infinite sets. Previous research data indicate that students usually do not apply one-to-one correspondence for such comparisons, they most commonly use considerations of inclusion (e.g., Borasi, 1985; Moreno & Waldegg, 1991), and their choices of criteria for comparisons of infinite sets are representation-dependent (e.g., Duval, 1983; Tsamir, in press). That is to say, in this study, unlike elsewhere, participants were asked to judge the acceptability of given criteria and not to come up with their own criteria for comparison. In reaction participants exhibited a non-negligible tendency to accept the one-to-one correspondence criterion. However, participants who had not taken a set theory course, and even a substantial number of those who had, also claimed that inclusion or single infinity were acceptable. Moreover, the N-ST participants tended to view inclusion as the most applicable, and interestingly, a number of C-ST participants referred to powers to reject one-to-one correspondence as a valid criterion, failing to grasp that one-to-one correspondence is the notion underlying powers.

Practical considerations played an important role in participants' decisions whether to accept or reject a specific criterion. Moreover, similar practical considerations led different participants to different conclusions. For example, the fact that sets are not necessarily inclusive led several participants to accept the inclusion criterion, claiming that "It is not always possible to use inclusion, but this is a criterion which is easy to use, so it should be used when possible." Other participants, however, quoted the same fact to reject this criterion. They explained that "Sets are not always inclusive, so this is not a good criterion." Furthermore, all participants tended to examine each problem in isolation, ignoring the contradictions which arise when accepting more than one of the above mentioned criteria as being valid. Consistency was only rarely mentioned as the consideration for deciding whether or not to accept a suggested criterion and as a means to determine validity in mathematics (see also Tsamir & Tirosh, 1994; Tsamir, 1999).

After having been presented in Stage II with illustrations of the use of one-to-one correspondence as a criterion for comparing infinite sets, at Stage III there was an increase among all participants in the acceptance of one-to-one correspondence. However, the rate of N-ST and even C-ST participants who accepted inclusion was unchanged, while the rate of acceptance of single infinity rose. It seems that while familiarizing the N-ST with the one-to-one correspondence criterion and reminding the C-ST participants of it, strengthens their tendency to accept this criterion, it does not weaken their tendency to accept other criteria. These findings support Fischbein's claim that

> new, correct intuitions do not simply replace primitive, incorrect ones. Primary intuitions are usually so resistant that they may coexist with new, superior, scientifically acceptable ones. That situation very often generates inconsistencies in the student's reactions depending on the nature of the problem. (1987, p. 213)

The findings also indicate that, as could be expected, participants who had studied Cantorian set theory exhibited a significantly higher tendency to accept one-to-one correspondence than participants who had not taken such a course. These findings were yielded by both Stages I and II—when asked to state whether one-to-one correspondence would be a suitable criterion for the comparison of infinite sets, and when asked to judge illustrated solutions to specific comparison tasks.

However, even after studying set theory, participants still failed to grasp one of its key aspects, that is, that the use of more than one of the above mentioned criteria for comparing infinite sets will eventually lead to contradiction. None of the set theory course graduates stated, for instance, that the use of any of the suggested criteria is acceptable under the condition that it is the only criterion used for comparing infinite sets. Nobody mentioned that as long as one criterion is applied, consistency is preserved and the mathematical theory is valid, nor did anybody argue that one-to-one correspondence was the (only) criterion preferred by Cantor, and thus the only criterion to be used within the Cantorian set theory.

As mentioned before, both young children and students fail to understand the categorical conclusion of equivalence evolving from one-to-one correspondence relationship between two given sets. Both confuse the significant role of one-to-one correspondence with that of other, irrelevant criteria. Young children relate to perceptual factors, such as the spatial arrangement of the elements in the two sets, and claim, when the sets have different spatial arrangement, that they are not equivalent. Prospective teachers also confuse the conclusive information derived from given one-to-one correspondence relationships between infinite sets, with practical considerations. They allowed interchangeable use of different criteria, such as inclusion and one-to-one correspondence, for the comparison of infinite sets, the use of which leads to contradictory results.

Piaget described how, gradually, with age, children develop a sense that "something is wrong" and exhibit "intermediary reactions" of systematic conflict, accompanied by different attempts to reconcile the incompatibilities (Piaget, 1941/1969, p. 32). Eventually, the correspondence shifts spontaneously from being held in check by perceptual or mere global factors between configurations into truly quantifying correspondence, which issues in necessary equivalence and thus in cardinal invariance (p. 41). Returning to the question posed in the introduction of this article: Is this the case for infinite sets as well? Are difficulties and contradictions related to students' grasp of the cardinality of infinite sets also resolved with age?

The findings related to the N-ST as well as the C-ST participants indicate that the answer is: "No." There is no such spontaneous development in students' grasp of the cardinality of infinite sets. Therefore, it seems that Fischbein's assertion, about the relationship between primary and secondary intuitions is valid, at least for the comparisons of infinite sets:

> One has to bear in mind that intuitively based conceptions cannot be eliminated simply by mere verbal explanations. . . . The development of new . . . mathematical and scientific intuition implies, then, didactical situations in which the student is asked to evaluate, to conjecture, to predict, to devise and check solutions. . . . It is certain that mathematics education cannot be successfully achieved by simply bypassing the intuitive obstacles through purely formal teaching. (Fischbein, 1987, pp. 38, 213)

Indeed, the present findings suggest that a merely formal presentation of Cantorian set theory, which tells students what to do (e.g., one-to-one correspondence is acceptable), instead of explicitly relating to their primary ideas, may not lessen participants' intuitively based conceptions. That is, "showing students the right way" was not enough. It did not decrease their tendency to accept incompatible, intuitive ideas when comparing infinite sets. Rather, students' primary intuitions should be taken into consideration when planning instruction. Students should be made aware of the pitfalls of their tendency to view all infinities as equal and of their tendency to use considerations of inclusion. For example, successful experiences of applying the cognitive-conflict teaching approach in teaching different mathematical topics, reported in the literature should be considered (Swan, 1983; Tirosh & Graeber, 1990). Moreover, students should understand what *not* to do and *why* (e.g., single infinity is not acceptable), and that using more than one of these criteria to compare infinite sets leads to contradiction.

One way to go about this is by creating a sequence of comparison tasks that advances from the comparison of finite to that of infinite sets, highlighting the crucial differences between the two types of comparison (see, for instance, Appendix, Tsamir, 1999). A different intervention can apply research findings regarding students' reactions to different representations of the same pair of infinite sets, in designing a sequence of comparison tasks aimed, for instance, to trigger cognitive conflict (see, for instance, Tsamir, in press; Tsamir & Tirosh, 1999). Another possibility is to promote the students' awareness of their intuitive ideas, as was found beneficial in other cases (Tirosh, 1991; Tsamir, 2002; Tsamir & Dreyfus, 2002).[4]

[4]Another suggestion, made by one of the reviewers, is to make connections between the notion of function and the notion of one-to-one correspondence used for the comparison of infinite sets.

Clearly, there is a need for further research in order to investigate the impact of the various interventions on students' performance with infinite sets. Prospective teachers, however, who are facing a future of mathematics teaching should be made aware of their own intuitive ideas and of the need for consistency and its link to formal knowledge. Sharing with them the rationale underlying activities designed to make them aware of their primary intuitions, and the constant need to "stand guard" and criticize their performance, may serve to improve their mathematical knowledge today and to make them better teachers in the future.

References

Borasi, R. (1985). Errors in the enumeration of infinite sets. *Focus on Learning Problems in Mathematics*, *7*(3 & 4), 77–89.

Cantor, G. (1885/1950). *Contribution to the foundation of the theory of transfinite numbers.* USA: Dover Publication.

Duval, R. (1983). L'obstacle du d/edoublement des objects math/ematiques. *Educational Studies in Mathematics*, *14*, 385–414.

Fischbein, E. (1987). *Intuition in science and mathematics.* Dordrecht, Holland: Reidel.

Fischbein, E. (1993). The interaction between the formal and the algorithmic and the intuitive components in a mathematical activity. In R. Biehler, R. Scholz, R. Straser, & B. Winkelmann (Eds.), *Didactics of mathematics as a scientific discipline.* Kluwer.

Fischbein, E., Tirosh, D., & Melamed, U. (1981). Is it possible to measure the intuitive acceptance of a mathematical statement? *Educational Studies in Mathematics*, *12*, 491–512.

Fraenkel, A. (1953). *An introduction to mathematics, problems and methods of the new mathematics.* Tel-Aviv: (in Hebrew) Massada.

Hilbert, D. (1924/1989). On the infinite. In P. Benacerraf & H. Putman (Eds.), *Philosophy of mathematics* (pp. 183–201). New York, USA: Cambridge University Press.

Moreno, L. E., & Waldegg, G. (1991). The conceptual evolution of actual mathematical infinity. *Educational Studies in Mathematics*, *22*, 211–231.

Piaget, J. (1941/1969). *The child's conception of number.* London: Rutledge & Kegan Paul.

Russell, B. (1903/1950). *The principles of mathematics.* London: George Allen & Unwin.

Russell, B. (1916/1980). Mathematics and the metaphysicians. In J. R. Newman (Ed.), *The world of mathematics* (Vol. 3, pp. 1551–1564). Washington, USA: Tempus Books.

Sierpinska, A., & Viwegier, M. (1989). How and when attitudes towards mathematics & infinity become constituted into obstacles on students? *Proceedings of the 13th Annual Meeting for the Psychology of Mathematics Education*, *3*, 166–173.

Swan, M. (1983). *Teaching decimal place value: A comparative study of "conflict" and "positive only" approaches.* Nottingham: University of Nottingham, Shell Centre for Mathematical Education.

Tirosh, D. (1991). The role of students' intuitions of infinity in teaching the cantorial theory. In D. Tall (Ed.), *Advanced mathematical thinking* (pp. 199–214). Dordrecht, Holland: Kluwer.

Tirosh, D., & Graeber, O. (1990). Evoking cognitive conflict to explore preservice teachers' thinking about division. *Journal for Research in Mathematics Edu cation*, *21*(2), 98–108.

Tirosh, D., & Tsamir, P. (1996). The role of representations in students' intuitive thinking about infinity. *Journal of Mathematical Education in Science and Technology*, *27*(1), 33–40.

Tsamir, P. (1999). The transition from comparison of finite to the comparison of infinite sets: Teaching prospective teachers. *Educational Studies in Mathematics*, *38*, 209–234.

Tsamir, P. (2002). When the "same" is not perceived as such: The case of infinite sets. *Educational Studies in Mathematics*, *48*, 289–307.

Tsamir, P. (in press). From "easy" to "difficult" or vice versa: The case of infinite sets. *Focus on Learning Problems in Mathematics.*

Tsamir, P., & Dreyfus, T. (2002). Comparing infinite sets—A process of abstraction: The case of Ben. *Journal of Mathematical Behavior*, *21*, 1–23.

Tsamir, P., & Tirosh, D. (1992). Students' awareness of inconsistent ideas about actual infinity. *Proceedings of the 16th Annual Meeting for the Psychology of Mathematics Education*, *3*, 90–97.

Tsamir, P., & Tirosh, D. (1994). Comparing infinite sets: intuitions and representations. *Proceedings of the 16th Annual Meeting for the Psychology of Mathematics Education*, *4*, 345–352.

Tsamir, P., & Tirosh, D. (1999). Consistency and representations: The case of actual infinity. *Journal for Research in Mathematics Education*, *30*(2), 213–219.

Appendix

Assignment: From Finite to Infinite Sets. This assignment comprises two parts. The first part deals with comparisons of finite sets and the following part is devoted to the comparisons of infinite sets. The main aims of the assignment are first to promote students' awareness of the different methods they apply when comparing finite sets, and of the similar methods they use in comparing infinite sets. Then, to draw their attention to the crucial issue of consistency, which is preserved when interchangeably using methods to compare finite sets, but is violated when interchangeably using methods to compare infinite sets.

The assignments may serve as a springboard for promoting students' awareness of the influence of their "finite experience" on their performance with infinite sets, and of the consequent contradictions that arise. The conclusion should be that when comparing infinite sets one single method should be used for all comparisons. One single method should be applied consistently, but this method can be selected from a number of valid ones. Discussion may deal with existent primary intuitions and their influence on our mathematical performance, the need to preserve consistency, and the important role of formal, mathematical justifications. In Part 1 the students are encouraged to use various available methods for comparing finite sets. The first task in this part is designed to evoke ideas of "counting," "one-to-one correspondence" and "inclusion." Students tend to respond to the problems by using different methods, usually, by using one criterion per one problem. The second task acquires them to apply various methods to each of the problems. In their reflection it is concluded that it is not always possible to apply all methods; however, when more than one method is applicable the results reached by all methods will be identical.

In Part 2 the students are encouraged to use various available methods for comparing infinite sets. Again, according to the literature and to the findings of my research, students tend to respond to each problem with a single method, and different problems trigger different methods.

They, then, are asked to apply all possible methods to each of the problems. This part is designed to lead the students to conclude that it is impossible to preserve consistency when more than one of the methods is applied. The application of different methods for comparing infinite sets eventually leads to contradiction.

Each part ends with a reflective section and a summary of both parts is done in the final, concluding section. The assignment is presented here in the sequence in which it is given to students.

Part 1: Comparing finite sets

Task 1.a. Compare the number of elements in the following pairs of finite sets:

Problem 1
Given: $A = \{5, 10, 15, 20, 25, 30\}$ $B = \{10, 20, 30\}$
Is the number of elements in set A equal to the number of elements in set B? Yes / No
How did you reach this conclusion? ______________________________

Problem 2
Given: $N = \{1, 2, 3, 4, 5\}$ $E = \{2, 4, 6, 8, 10\}$
Is the number of elements in set N equal to the number of elements in set E? Yes / No
How did you reach this conclusion? ______________________________

Problem 3
A dinner table is set. Each person is served a plate, a glass, a knife and two forks.
$F = \{\text{The forks}\}$ $K = \{\text{The knives}\}$ $G = \{\text{The glasses}\}$
Is the number of elements in set F equal to the number of elements in set K?
Yes / No How did you reach this conclusion? ______________________________

Is the number of elements in set G equal to the number of elements in set K? Yes / No
How did you reach this conclusion? ______________________________

Problem 4
Given: $V = \{1, 2, 3, 4, 5 \ldots 998, 999, 1000\}$ $J = \{10, 20, 30 \ldots 980, 990, 1000\}$
Is the number of elements in set V equal to the number of elements in set J? Yes / No
How did you reach this conclusion? ______________________________

Problem 5
Given: $B = \{3, 5, 7, 9, \ldots 1977\}$ $Y = \{3^2, 5^2, 7^2, 9^2, \ldots 1977^2\}$
Is the number of elements in set B equal to the number of elements in set Y? Yes / No
How did you reach this conclusion? ______________________________

Let's Reflect: What were the criteria that you used to compare the given sets?

Task 1.b. Try to apply the three criteria (one-to-one correspondence, inclusion, counting) to each of the previous problems.

Sample Problem: Problem 1
Given the sets: $A = \{5, 10, 15, 20, 25, 30\}$ $B = \{10, 20, 30\}$
Is the number of elements in set A equal to the number of elements in set B?

In order to answer this question—

Is "one-to-one correspondence" applicable? Yes / No
 If your answer is Yes— Use this criterion to solve the problem:
 Is the number of elements in set A equal to number in set B? Yes / No

Is "inclusion" applicable? Yes / No
 If your answer is Yes— Use this method to solve the problem:
 Is the number of elements in set A equal to number in set B? Yes / No

Is "counting" applicable? Yes / No
 If your answer is Yes— Use this method to solve the problem:
 Is the number of elements in set A equal to number in set B? Yes / No

Let's Reflect: Do we reach different conclusions by applying different criteria for the comparison of the same pair of finite sets?

Part 2: Comparing infinite sets

Task 2.a. Compare the number of elements in the following pairs of infinite sets:

Problem A
Given $T = \{1, 2, 3, 4, 5, \dots\}$ $P = \{1^2, 2^2, 3^2, 4^2, 5^2, \dots\}$
The number of elements in sets T and B is equal / not equal. Explain:

Problem B
Given $F = \{5, 10, 15, 20, 25, 30, \dots\}$ $K = \{10, 20, 30, \dots\}$
The number of elements in sets F and K is equal / not equal. Explain:

Problem C
Given $E = \{2, 4, 6, 8, 10, \dots\}$ $S = \{1, 2, 3, 4, 5, \dots\}$
The number of elements in sets S and R is equal / not equal. Explain:

Let's Reflect: What are the criteria that you used to compare the given sets?

Task 2.b. Try to apply all three criteria (one-to-one correspondence, inclusion and all infinite sets are equal) to each of the previous problems.

Sample Problem: Problem A
Given $T = \{1, 2, 3, 4, 5, \dots\}$ $P = \{1^2, 2^2, 3^2, 4^2, 5^2 \dots\}$
The number of elements in sets P and T is equal / not equal.

In order to answer this question—

Is "one-to-one correspondence" applicable? Yes / No
If your answer is Yes— Use this criterion to solve the problem:
Is the number of elements in set T equal to number in set P? Yes / No

Is "inclusion" applicable? Yes / No
If your answer is Yes— Use this method to solve the problem:
Is the number of elements in set T equal to number in set P? Yes / No

Is "all infinities are equal" applicable? Yes / No
If your answer is Yes— Use this method to solve the problem:
Is the number of elements in set T equal to the number in set P? Yes / No

Let's Reflect: Do we reach different conclusions by applying different criteria for the comparison of the same pair of infinite sets?

Summing Up

Considering the various methods that were used to compare finite sets and those used to compare infinite sets, what happens in the extension from finite to infinite sets?

School of Education, Tel Aviv University, Ramat Aviv, Tel Aviv 69978, ISRAEL
E-mail address: pessia@post.tau.ac.il

CBMS Issues in Mathematics Education
Volume **12**, 2003

Student Performance and Attitudes in Courses Based on APOS Theory and the ACE Teaching Cycle

Kirk Weller, Julie M. Clark, Ed Dubinsky, Sergio Loch, Michael A. McDonald, and Robert R. Merkovsky

Abstract. Over the last several years a number of mathematics education researchers have applied a particular research framework to study student learning of various topics in the undergraduate mathematics curriculum. Included in this research framework is the development and implementation of instructional treatments based upon APOS Theory, an extension of Piaget's theory of reflective abstraction to the undergraduate mathematics curriculum, and the ACE Teaching Cycle, a pedagogical approach that encourages active student learning. Previous works by these researchers have focused on the nature of students' understandings using data that measured students' performances on various mathematical tasks. For the most part, such data were generated by small samples of students whose selection was not based upon the use of rigorous sampling techniques. The purpose of this paper is to examine such data collectively as a means of gauging the overall effectiveness of instruction based upon APOS Theory and the ACE Teaching Cycle. This report, based upon the data gathered in fourteen previous studies in the areas of calculus, abstract algebra, concept of function, quantification, induction, and the affective domain, paints a picture that suggests that instruction based upon APOS Theory may be an effective tool in helping students to learn mathematical concepts.

Some mathematics education researchers believe that one can best study how students develop their understandings of mathematical concepts by adopting a single research paradigm. One such group of researchers, RUMEC,[1] has chosen to use a research framework that consists of theory, an approach to the design and implementation of instruction, and a method for collecting and analyzing data. The theoretical perspective, called APOS Theory, is an extension of Piaget's theory of reflective abstraction applied to the undergraduate mathematics curriculum. APOS is an acronym for **A**ction, **P**rocess, **O**bject, and **S**chema, the mental constructions that students are likely to make in formulating their understandings of mathematical concepts. The instructional approach, referred to as the ACE Teaching Cycle, attempts to foster the development of certain mental constructions by

[1]RUMEC stands for Research in Undergraduate Mathematics Education Community. More information regarding the work of RUMEC can be found at `http://www.cs.gsu.edu/~rumec`.

de-emphasizing the lecture method in favor of problem solving, cooperative learning, and the use of a mathematical programming language. The research method is primarily qualitative. All of these are described in the next section and discussed in detail in Asiala et al. (1996).

Up to this point, the aim of nearly every study based on this framework has been to reveal the nature of students' understandings rather than to provide statistical comparisons of student mathematical performance. However, one important aspect of the research framework is to describe the degree to which students successfully complete various mathematical tasks after having completed the instructional treatment. Because the data collected in previous studies were generated from small samples in which the students were selected somewhat informally, no attempt was made to analyze them in any systematic manner. To address this issue, the authors of this paper have analyzed the numerical results of students' mathematical performances and attitudes from a number of completed studies, with the goal of trying to determine what the data suggest regarding the effectiveness of instruction based upon APOS Theory and the ACE Teaching Cycle.

Fourteen studies were selected for this analysis: four focusing on calculus (Asiala, Cottrill, Dubinsky, & Schwingendorf, 1997; Baker, Cooley, & Trigueros, 2000; Clark et al., 1997; McDonald, Mathews, & Strobel, 2000), three on abstract algebra (Asiala, Brown, Kleiman, & Mathews, 1998; Asiala, Dubinsky, Mathews, Morics, & Oktac, 1997; Brown, DeVries, Dubinsky, & Thomas, 1997), one on the concept of function (Breidenbach, Dubinsky, Hawks, & Nichols, 1992), two on quantification (Dubinsky, 1997; Dubinsky, Litman, Morics, & Oktac, n. d.), one on induction (Dubinsky, 1989), two on the affective domain (Asiala & Dubinsky, 1999; Clark, Hemenway, John, Tolias, & Vakil, 1999), and one longitudinal study (Kuhn, McCabe, & Schwingendorf, 2000). Each of the studies was selected because it contained numerical data on students' performances on specific mathematical tasks and/or on their attitudes toward mathematics. With the exception of Asiala and Dubinsky (1999) and Dubinsky et al. (n. d.), each study has either been published or accepted for publication. Projects that are still in process or completed projects that do not report such data have not been included in this paper. It is important to note that the studies are not independent because in some cases multiple studies were generated from the same large data sets. These studies can be classified in the following ways:

- comparative studies in which the performance of students who received instruction using APOS Theory and the ACE Teaching Cycle is compared with the mathematical performance of students who completed traditional lecture/recitation courses;
- non-comparative studies measuring the performance of students who completed courses using APOS Theory and the ACE Teaching Cycle;
- studies of the level of cognitive development of students who completed courses based upon APOSTheory and the ACE Teaching Cycle or courses using a traditional lecture/recitation model;
- comparisons of student attitudes and the long-term impact of courses based upon APOS Theory and the ACE Teaching Cycle to that of students who completed traditional lecture/recitation courses.

When taken collectively, these studies suggest that instruction based upon APOS Theory and using the ACE Teaching Cycle is effective, and that APOS

Theory is a promising tool for describing how students construct mathematical concepts. In particular, these studies will provide evidence that:

- within a comparative context, students who received instruction using curriculum based upon APOS Theory and the ACE Teaching Cycle performed at least as well, and in most cases better, than students who completed traditional lecture/recitation courses;
- with respect to non-comparative data, the level of student performance was higher than that observed by the authors in their use of traditional lecture/recitation models of instruction;
- with respect to cognitive development, students who received instruction using APOS-based curricular materials achieved a reasonably mature understanding of the concepts being presented;
- with respect to the affective domain, instruction using APOS curricular materials and delivered via the ACE Teaching Cycle had a positive effect upon students' interest in and enjoyment of mathematics.

The balance of this paper consists of a brief description of APOS Theory and the ACE Teaching Cycle followed by six sections presenting the results of student performance in the subject areas of calculus, abstract algebra, concept of function, quantification, mathematical induction, and the affective domain. The paper ends with sections presenting statistics on student performance in courses subsequent to instruction based upon APOS Theory, as well as concluding remarks and observations.

Each section that presents the results of student performance data contains a brief description of the groups of students whose performances were recorded and a summary of the instruments and the methods used to generate and to analyze the data. All performance results are presented in tabular form.

1. The Research and Development Framework

In this section we give a brief description of the components of the research and development paradigm used by the researchers in each of the fourteen studies. In Section 2.1, we describe APOS Theory. In Section 2.2, we outline the components of the instructional design, its implementation, and the method used for collecting and analyzing data.

1.1. APOS Theory. APOS Theory, a constructivist theory rooted in the work of Jean Piaget (1965/1966) on the concept of reflective abstraction, has been developed as one mechanism by which researchers can examine and describe students' advanced mathematical thinking. The theory has been built from the hypothesis that an individual's mathematical knowledge consists in her or his tendency to deal with a mathematical situation by constructing mental actions, processes, and objects, and organizing them into schemas to make sense of the situation. APOS is an acronym for **A**ction, **P**rocess, **O**bject, and **S**chema—a description of the mental activities and mental constructions that students might tend to make in formulating their understandings of mathematical concepts.

An *action* is any transformation of objects (physical or mental) to obtain other objects. It is perceived by an individual to be essentially external and, as such, requires, either explicitly communicated or from memory, step-by-step instructions on how to perform the transformation. A student is limited to an *action conception* of a given concept if her or his depth of understanding is limited to performing actions relative to that concept.

As an individual repeats an action, he or she may reflect upon it until it becomes *interiorized* into a mental *process*, which the individual perceives as being under her or his control. In contrast to an action conception, a *process conception* is characterized by an individual's ability to describe, to reflect upon, or to reverse the steps of a transformation without actually having to perform the steps explicitly. An individual is limited to a process conception of a concept if her or his depth of understanding is limited to thinking about the concept exclusively within a procedural context.

When the individual becomes aware of a process as a totality and realizes that he or she can create transformations that can act upon the given process, the individual has *encapsulated* her or his process conception into a mental *object*. An individual has an object conception of a concept if he or she can *de-encapsulate* the object conception back to its underlying process and construct transformations that can be applied to the concept.

A *schema* for a certain mathematical concept or topic is an individual's collection of actions, processes, objects, and other schema which are linked, consciously or unconsciously, into a coherent mental framework which the individual may access when faced with a mathematical problem situation involving that concept or topic. The coherence of a schema is characterized by an individual's ability to identify which items fit within the scope of the schema. The formation of a schema may progress through three levels of development: *intra*, *inter*, and *trans* (Clark et al., 1997; Piaget & Garcia, 1983/1989). An individual at the intra stage may have constructed a number of the components that would be contained in the schema of a given concept, but he or she may be unable to recognize the relationships that exist between those components. An individual at the inter stage begins to recognize relationships between related mental constructs within the schema, but he or she does not yet conceive the manner in which such constructs form a single cognitive entity. An individual reaches the trans stage when he or she has constructed a coherent structure through which the relationships between the components constituting the schema are understood as a single entity.

APOS Theory has been used in designing curricula and instructional treatments that may enhance students' mathematical understandings and performances. In particular, the theory was utilized in designing the courses that form the basis of this report.

1.2. The Research and Development Framework. The research framework has three components. Researchers begin their inquiry by constructing an initial *genetic decomposition*, an explicit description of the constructions students are likely to make in formulating their understandings of a given concept. The initial genetic decomposition motivates the design of an instructional treatment referred to as the *ACE Teaching Cycle*, which itself consists of three components, which, in order, are *activities*, *class discussion*, and *exercises*. The activities, which form the heart of the instructional treatment, have typically involved the use of a mathematical programming language, which students use to write short computer programs designed to foster development of the proposed mental constructions (Dubinsky, 1995). The class discussion is led by the instructor and is designed to give students an opportunity to reflect on the work done in the computer lab. The exercises are relatively traditional homework problems that serve the purposes of reinforcing the ideas that the students have constructed, stimulating thought about situations that will be studied later, and providing the students with occasions to use the mathematics they have learned. During and/or following implementation of the instructional treatment, researchers gather and analyze the data in the context of the theory. The data analysis may suggest a refinement of either the specific genetic decomposition, the instructional design, or the underlying theory. Researchers test subsequent revisions of the genetic decomposition and the instructional treatment by cycling through the components of the framework. This cyclical process continues until researchers achieve a level of equilibrium in their conceptualization of students' interactions with the concept being studied.

The researchers employ both qualitative and quantitative methods of data analysis. Data for the fourteen studies discussed in this paper were collected from written instruments and in-depth interviews, which were subsequently recorded and transcribed.

In addition to qualitative methods of categorization and interrelation, researchers seek to explain differences in students' mathematical performances in terms of the specific mental constructions of actions, processes, objects and/or schemas that students make, are in the process of making, or fail to make. The specific mental constructions that arise from such an analysis may motivate the researchers to modify the initial or existing genetic decomposition.

For a more detailed discussion of APOS Theory, the ACE Teaching Cycle, and the research framework, the reader should consult Asiala et al. (1996) and Clark et al. (2000).

2. Student Performance: Calculus

2.1. Related Papers and Student Characteristics. Four studies examined various aspects of student performance in calculus.

- "The Development of Students' Graphical Understanding of the Derivative" (Asiala et al., 1997) explored students' understandings of the graph of a function and its derivative.
- "A Calculus Graphing Schema" (Baker et al., 2000) considered students' ability to construct the graph of a function when given only conditions involving continuity, derivatives, and asymptotes.
- "Constructing a Schema: The Case of the Chain Rule" (Clark et al., 1997) investigated the nature and development of students' conceptions of the chain rule.
- "Understanding Sequences: A Tale of Two Objects" (McDonald et al., 2000) examined the mental constructions students might make in developing their conceptions of sequences.

The data used in Asiala et al. (1997), Baker et al. (2000), and Clark et al. (1997) were collected in interviews conducted with one group of 41 students who had completed at least two semesters of univariate calculus at a large midwestern university between the fall of 1988 and the spring of 1990.

Of these students, 17 completed the Calculus, Concepts, Computers, and Cooperative Learning sequence (C^4L), a calculus curriculum development project based upon APOS Theory and delivered via the ACE Teaching Cycle (Dubinsky & Schwingendorf, 1991; Dubinsky, Schwingendorf, & Mathews, 1995). We will call this the experimental group (EXP). The remaining 24 students completed a traditional calculus sequence. By traditional, we mean a lecture-oriented mode of delivery, use of a non-reform text, and evaluation based almost exclusively on in-class tests and final examinations. Students having completed such a sequence will be referred to as the traditional group (TRAD).

The analyses conducted in Asiala et al. (1997), Baker et al. (2000), and Clark et al. (1997) were based upon students' responses to six items. Two of the items were designed to ascertain students' understandings of the graph of a function. Analysis of students' responses to these items, the focus of Asiala et al. (1997) and Baker et al. (2000), is considered in Section 3.2. The other four questions, all of which involved the chain rule, were the instruments used in Clark et al. (1997). The comparative performance statistics for this study are presented in Section 3.3.

Of the 41 students who were subjects for Asiala et al. (1997), Baker et al. (2000), and Clark et al. (1997), 21 of these students were also subjects for McDonald et al. (2000). For this study, 15 of the 21 students completed the C^4L sequence, and the remaining 6 completed traditional courses. The authors analyzed students' responses to 14 questions involving sequences. Student performance data for this study are considered in Section 3.4.

Although the traditional and C^4L courses employed different pedagogical strategies, the major conceptual ideas studied in all four papers being considered here were central to the curriculum of both the C^4L and traditional sequences. Generally speaking, the instructors for the experimental sections were involved in some way in the research. The traditional courses were taught by a variety of faculty, some of whom were award-winning teachers. For each study, the data were transcripts of audio-taped interviews administered to the students at various times ranging from one-half to three semesters after they had finished their study of univariate calculus. The students for these studies were not self-selected, although little attempt was made to randomize their characteristics. A review of their overall academic records and a comparison of their Predicted Grade Point Averages (a tool developed by the university to predict a prospective student's likely college GPA) suggest that the ability levels of the two groups were quite similar (Kuhn et al., 2000).

2.2. Graphical Understanding of the Derivative. Students' understandings of the graph of a function and its derivative were the focus of Asiala et al. (1997) and Baker et al. (2000). The instruments used in these studies are discussed in Section 3.2.1. The performance results from Asiala et al. (1997) are presented in Section 3.2.2. The results from Baker et al. (2000) are given in Section 3.2.3.

2.2.1. *Instruments.* The comparative performance data reported in Asiala et al. (1997) and Baker et al. (2000) were based upon students' responses to the two interview questions given below. Student responses to both questions were used in Asiala et al. (1997), and data gathered from the second question were the subject of Baker et al. (2000).

(1) Suppose that the line L is tangent to the graph of the function f at the point $(5, 4)$ as indicated in Figure 1. Find $f(5)$ and $f'(5)$. Explain how you obtained your answer.

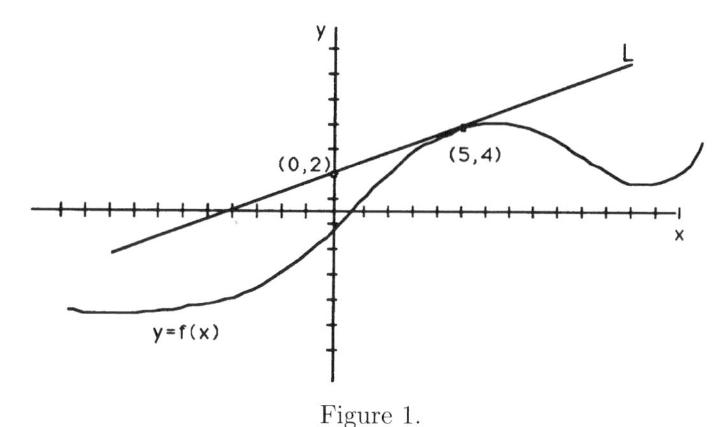

Figure 1.

(2) Sketch a graph of a function h which satisfies the following conditions:
h is continuous
$h(0) = 2$, $h'(-2) = h'(3) = 0$, and $\lim_{x \to 0} h'(x) = \infty$,
$h'(x) > 0$ when $-4 < x < -2$, and when $-2 < x < 3$,
$h'(x) < 0$ when $x < -4$, and when $x > 3$,
$h''(x) < 0$ when $x < -4$, when $-4 < x < -2$, and when $0 < x < 5$,
$h''(x) > 0$ when $-2 < x < 0$, and when $x > 5$,
$\lim_{x \to -\infty} h(x) = \infty$ and $\lim_{x \to \infty} h(x) = -2$.

2.2.2. *Comparative Data: Students' Graphical Understanding of the Derivative.* Because the nature of the interview process did not lend itself to the collection of numerical data on specific mathematical tasks, the authors used interview transcripts, together with APOS Theory, to develop two indicators that measured students' general understandings about functions and function notation (Asiala et al., 1997). For each indicator, the percentage of students having met each criterion from the C^4L (EXP) group is contrasted with that of students having completed traditional calculus sequences (TRAD).

Table 1. General Understanding about Functions and Notation

Indicator	EXP		TRAD	
	%	$N = 17$	%	$N = 24$
The student provided evidence of a process conception of function, both in understanding $f(x)$ notation and in the ability to work with the graph of a function in the absence of an algebraic expression.	100	17	67	16
Given the point $(a, f(a))$, the student understood that $f'(a)$ gives the slope of the tangent at $x = a$. In the absence of the algebraic expression, the student was able to think about and work with the derivative using only graphical information.	76	13	50	12

In addition to this general understanding, the authors developed eight additional criteria, four to measure the graphical understanding of a function, and four to measure the graphical understanding of the derivative. In determining the degree to which a student satisfied the condition of a given criterion, the authors devised a 3-point rating system. If a student appeared to demonstrate complete understanding, the authors assigned a 3 rating. If a student appeared to have grasped the main ideas, but was prompted and/or made specific errors, the authors assigned a 2 rating. If the student displayed little or no understanding, the authors assigned a 1 rating. The rating given to a student for a particular criterion was based upon a consensus decision. No rating was assigned until all of the authors reached agreement. Since each student's ratings on all eight criteria were recorded together, the authors knew how each student performed on all eight criteria. The specific criteria, the rating system, and the comparative data for the experimental (EXP) and traditional (TRAD) groups are presented in Table 2.

Table 2. Graphical Understanding of a Function and its Derivative

Indicator	Rating	EXP		TRAD	
		%	$N = 17$	%	$N = 24$
Appeared to understand $y = f(x)$ notation as presented on the graph.	3	94	16	54	13
	2	6	1	17	4
	1	0	0	29	7
Appeared to understand the functional notation in general.	3	94	16	75	18
	2	6	1	17	4
	1	0	0	8	2
Was able to deal with the function based only on graphical information, without explicit use of expression.	3	100	17	71	17
	2	0	0	0	0
	1	0	0	29	7
Was successful in drawing a graph of a function from specific information given about values of the function and its derivative.	3	65	11	29	7
	2	29	5	38	9
	1	6	1	33	8
Appeared to understand that the value of $f'(x)$ is the slope of the tangent to the graph at $(x, f(x))$.	3	71	12	46	11
	2	18	3	13	3
	1	12	2	42	10
Was able to deal with the derivative of the function based only on graphical information, without using expressions.	3	71	12	50	12
	2	29	5	0	0
	1	0	0	50	12
Appeared to understand that, as the value of the derivative approaches infinity, the slope of the graph of the function grows without bound, yielding a vertical asymptote.	3	82	14	63	15
	2	0	0	4	1
	1	18	3	33	8
Appeared to understand how to use the derivative to determine intervals of monotonicity for the function.	3	100	17	71	17
	2	0	0	0	0
	1	0	0	29	7
AVERAGE RATING PER INDICATOR			2.8		2.3

The results suggest that, when compared with traditional instruction, instruction based upon APOS Theory may yield the following benefits:

- Higher likelihood of students achieving a process-level conception of function;
- Increased ability of students working effectively with graphical information in the absence of algebraic information;
- Increased probability of students exhibiting high achievement in graphical understanding.

2.2.3. *Comparative Data: A Calculus Graphing Schema.* As noted in Section 2.1, APOS Theory incorporates a triad mechanism to explain schema development. Consideration of the Chain Rule in Clark et al. (1997) served as a major catalyst for expansion of the original single-schema conception to levels, which are referred to as the *intra*, *inter*, and *trans* stages of schema development. Baker et al. (2000) used this expanded conception of schema to analyze students' responses to Question 2 (described in Section 3.2.1). This involved creation of a double-triad mechanism, a means of determining the degree to which a student is able to coordinate graphical properties across contiguous intervals of the domain of a function. Each stage of the double-triad used in the analysis is briefly described below; for more detail, see Baker et al. (2000).

> *Intra-Property, Intra-Interval Level:* The student can produce a few isolated actions on the given properties of the function. The actions result from the relation of a single property to isolated intervals of the graph. There is no coordination across intervals nor among multiple conditions on a single interval.
>
> *Intra-Property, Inter-Interval Level:* The student is able to coordinate two or more, but not all, contiguous segments of the graph using one property.
>
> *Intra-Property, Trans-Interval Level:* The student is able to coordinate a single property consistently across the domain and to sketch a form of the graph that is consistent with the property being used.
>
> *Inter-Property, Intra-Interval Level:* The student can coordinate some, but not all, of the properties on one or more disjoint intervals. The student is not able to link contiguous intervals on the domain.
>
> *Inter-Property, Inter-Interval Level:* The student can coordinate some, but not all, properties across two or more, but not all, contiguous intervals of the domain.
>
> *Inter-Property, Trans-Interval Level:* The student can coordinate at least two, but not all, properties across the whole domain. The graph consistently reflects the properties being used.
>
> *Trans-Property, Intra-Interval Level:* The student can coordinate all properties on isolated intervals. No relationship between contiguous segments of the graph is achieved.
>
> *Trans-Property, Inter-Interval Level:* The student can coordinate all properties across some contiguous segments of the domain.

Trans-Property, Trans-Interval Level: The student can consistently coordinate all properties across the entire domain of the graph.

Table 3 shows the analysis of the experimental and the traditional students' responses to the graphing task. Each cell shows the percentage and the number of traditional (TRAD) and experimental (EXP) students assigned to each category of the double-triad. In analyzing each student's responses, the authors assigned each student to a category only after having negotiated a consensus result. Interval coordination is represented in the table column-wise. Graphical properties of a function are represented by the rows of the table.

Table 3. Schema Development Related to Graphical Understanding

	N	Intra	Inter	Trans
EXP	17			
Intra		18% (3)	12% (2)	0% (0)
Inter		0% (0)	12% (2)	18% (3)
Trans		0% (0)	12% (2)	29% (5)
TRAD	24			
Intra		13% (3)	21% (5)	13% (3)
Inter		8% (2)	8% (2)	17% (4)
Trans		0% (0)	8% (2)	13% (3)

In trying to make sense of Table 3, it may be considered that students whose responses were classified as being down and/or to the right of any given cell generally exhibited more fully developed graphical conceptions. For example, students who were deemed to be at a Trans(properties)/Trans(interval) level (lower right hand cell) tended to provide more sophisticated, complete, and correct responses than did students classified as being at the Intra(properties)/Intra(properties) (upper left hand cell) level. The statistics in this table, which seem to favor the experimental group, are consistent with the results reported in Asiala et al. (1997) in Section 3.2.2. Although based upon a somewhat different type of analysis, the results reported here add further credence to the conclusions listed at the end of Section 3.2.2.

2.3. Students' Understandings of the Chain Rule. Clark et al. (1997) analyzed students' responses to four differentiation tasks that required application of the Chain Rule. The authors judged students' responses according to the criteria given below:

S = successful (allowed for non-calculus computational errors only)
C = close to correct (right strategy/wrong derivative)
P = needed non-trivial prompt from interviewer
U = was unable to complete task even with interviewer prompt

For each question, the rating assigned to each student response was based on a consensus decision. No rating was assigned until all of the authors reached agreement. The authors in this paper have designated S and C ratings to be indicators of successful performance and a U rating to be a rating of unsuccessful performance. Table 4 gives the per task average for each rating level on the four items considered in the analysis in Clark et al. (1997).

(1) Differentiate $(1-4x^3)^2$ two different ways and compare the methods.
(2) Compute the derivative of $F(x) = \int_0^{\sin x} e^{t^2}\,dt$ and explain.

(3) Let A be a real number. Given that the following relation defines a function, $x\sqrt{y} + y\sqrt{x} = A$, find its derivative.

(4) A ladder A feet long is leaning against a wall, but sliding away from the wall at the rate of 4 ft/sec. Find a formula for the rate at which the top of the ladder is moving down the wall.

Table 4. Average Ratings for Chain Rule Questions

	EXP		TRAD		
Rating	%	N	%	N	Success Indicator
S or C	53	9	47	11	Successful
P	25	4	13	3	
U	22	4	40	10	Unsuccessful
Total		17		24	

Although the indicator of successful performance does not yield a sharp difference, the statistics presented here are consistent with those presented in the previous two sections in one important respect: the percentage of students having made little or no progress in the traditional group substantially exceeds that in the experimental group.

2.4. Students' Understandings of Sequences. McDonald et al. (2000) examined the mental constructions students might make in developing their conceptions of sequences. Their analysis used APOS Theory to determine students' levels of cognitive development on the basis of responses to 14 interview questions covering the following items:

- Define, compare, contrast, and give examples of sequences and series.
- Discuss the sequence $a_n = 1 + (-1)^n$.
- Discuss the sequence where each term is the average of the previous two terms.
- Explain the difference between sequences and functions.
- Define and give examples of sequences which are and are not monotone, bounded, or convergent.
- Draw a Venn diagram relating sequences which are, or are not, monotone increasing, bounded above, or convergent.
- Find $\lim\limits_{n\to\infty} \dfrac{2n^2-1}{1+5n^2}$ and then rigorously discuss what it means to say $\lim\limits_{n\to\infty} \dfrac{2n^2-1}{1+5n^2} = \dfrac{2}{5}$.
- Discuss the behavior of sequences of numbers such as $[0.9, 0.99, 0.999, \ldots]$ or $[2.7, 2.71, 2.718, 2.7182, 2.71828, 2.718281, \ldots]$.

McDonald et al.'s theoretical analysis involved coordination of two objects: SEQLIST, the encapsulation of the process of listing numbers explicitly to display a pattern, and SEQFUNC, the ability to see that a sequence is a function with positive integers as the domain. Each student's understanding was analyzed on the basis of whether he or she achieved process or object conceptions of SEQLIST and SEQFUNC, as well as her or his level of schema development. As with the other calculus studies, the authors determined a given student's level of progress only after having reached consensus. Before presenting the comparative statistics in Table 5, we give a brief description of each stage of schema development.

Intra Stage of Sequence Schema Development: Although there is a natural mathematical link between SEQLIST and SEQFUNC, since SEQLIST is the output of the SEQFUNC process, students treat the two constructs SEQLIST (object) and SEQFUNC (process or object) as being distinct and cognitively unrelated even though they may use both constructs in solving a particular mathematical problem.

Inter Stage of Sequence Schema Development: A student at the inter stage may demonstrate the existence of links between SEQFUNC and SEQLIST by being able to generate a list from a functional expression and by being able to write an algebraic expression after having examined patterns in a list. Students operating at this level of schema development preferentially apply SEQLIST or SEQFUNC, depending on the context or problem situation. In contrast to students at the intra stage, students at the inter stage may simultaneously be aware of the two constructions, but they may not be aware of their equivalence.

Trans Stage of Sequence Schema Development: Students at the trans stage of schema development have constructed both SEQLIST and SEQFUNC as cognitive objects. In addition, they have constructed strong and numerous connections between SEQLIST and SEQFUNC in such a manner that they are aware of these connections and subordinate each conceptualization to an overall concept of sequence.

Table 5. Levels of Development: Students' Conceptions of Sequences

APOS Level	EXP		TRAD	
	%	$N = 15$	%	$N = 6$
Object Conception of SEQLIST	100	15	100	6
Object Conception of SEQFUNC	67	10	33	2
LEVELS OF SCHEMA DEVELOPMENT:				
Trans Stage	40	6	17	1
Inter Stage	53	8	67	4
Transition from Intra to Inter Stage	7	1	17	1

The statistics presented here suggest that students who completed the C^{4}L courses tended to be more likely to have developed an object conception of a sequence as a function and to have coordinated successfully the two conceptions of sequence into a coherent, unified concept. These results are consistent with those reported in Sections 3.2.2, 3.2.3, and 3.3.

Given the variety of topics considered and the tendency of the comparative performance results to point in the direction of the experimental group, one could reasonably conclude that instruction employing APOS Theory and delivered via the ACE Teaching Cycle may result in a higher degree of student achievement and understanding in calculus than that which is often achieved using traditional methods.

3. Student Performance: Abstract Algebra

3.1. Related Papers and Student Characteristics. Three studies examined various aspects of student performance in abstract algebra.

- "The Development of Students' Understanding of Permutations and Symmetries" (Asiala et al., 1998) investigated how abstract algebra students might develop their understandings of permutations of a finite set and of symmetries of a regular polygon.
- "Student Understanding of Cosets, Normality, and Quotient Groups" (Asiala et al., 1997) reported on a study of the mental constructions that students might make in developing their understandings of cosets, normality, and quotient groups.
- "Learning Binary Operations, Groups, and Subgroups" (Brown et al., 1997) explored the manner in which students formulate their conceptualizations of binary operations, groups, and subgroups.

The participants for all three studies were drawn from a group of 51 students who completed at least one semester of an introductory abstract algebra course at a large midwestern university. Of these students, 31 completed a course using APOS Theory and the ACE Teaching Cycle. We will call this the experimental group (EXP). The remaining 20 students completed lecture-oriented courses at various times during the three semesters leading up to and including the semester in which the experimental course was given. In these courses, the instructors used traditional texts and evaluated student work almost exclusively through the use of in-class tests and final examinations. We will refer to this as the traditional group (TRAD). The students were not self-selected, although little attempt was made to randomize their characteristics. The instructors for the experimental courses included researchers involved with the three abstract algebra studies.

In Asiala et al. (1997) and Brown et al. (1997), both comparative and non-comparative data were considered. In Asiala et al. (1998), the analysis was based exclusively upon non-comparative data collected from the experimental group.

The latter data were gathered from two types of instruments administered exclusively to the 31 experimental students. The first instrument consisted of three written examinations given at various times throughout the semester. The second was a series of individual, audio-taped interviews conducted with 24 of the 31 experimental students immediately after the conclusion of the semester.

Comparative data in Asiala et al. (1997) and Brown et al. (1997) were gathered from a second audio-taped interview that was conducted with 17 of the experimental students in the semester immediately following completion of the APOS course and with all 20 traditional students one to three semesters after completion of their courses.

The non-comparative performance statistics from all three algebra papers are presented in Section 4.2. Comparative statistics related to students' understandings of groups, subgroups, normality, and cosets are considered in Section 4.3.

3.2. Non-Comparative Abstract Algebra Studies. Of the three written examinations, one was administered to students working in groups, while the other two were administered on an individual basis. The interviews were conducted with 24 of the 31 experimental students on an individual basis. In Table 6, under each topic, we present questions or descriptions of questions related to the given topic, together with performance results.

Table 6. Algebra Performance of EXP Students

Question	Performance
GROUPS AND BINARY OPERATIONS	
Items on Group Test	
Let $(G, *)$ be an abelian group, t a fixed element of G, and define a binary operation $\diamondsuit$ by $x \diamondsuit y = x * y * t^{-1}$, $x, y \in G$. Prove or disprove that $(G, \diamondsuit)$ is a group.	3/7 groups received full marks in proving that $(G, \diamondsuit)$ With partial credit, the average score was 89%.
Let $(G_1, *_1)$ and $(G_2, *_2)$ be two groups. Show that $G_1 \times G_2$, where $[a, b] * [c, d] = [a *_1 c, b *_2 d]$, is a group.	7/7 groups received full marks.
Items on Individual Test	
If R is a ring with identity, denote by R^* the set of units in R with the multiplication operation from the ring.	39% (12/31) either received full marks or made only minor errors.
Show that R^* forms a group.	With partial credit, the average score was 68%.
Let R be a ring and 0 the additive identity. Show that for all $x \in R$ it is the case that $x \cdot 0 = 0$.	All 31 (100%) students exhibited the ability to coordinate the two processes underlying the two ring operations. With partial credit, the average score was 77%.
SUBGROUPS	
Items on Group Test	
Prove or provide a counterexample: Every subgroup of an abelian group is abelian.	7/7 groups received full marks.
Find all subgroups of D_4.	4/7 groups were able to find all of the subgroups. The remaining 3 groups found all but 1 or 2 of the subgroups of order 4.
Find a subgroup of S_4 that is the same as S_3.	7/7 groups received full marks.
True/False Questions on Individual Test	
Given a closed subset of a group, you can be sure that the subset is a subgroup.	90% (28/31) correct responses
S_n is isomorphic to a subgroup of S_{n+2}.	71% (22/31) correct responses
A coset of a subgroup is always a subgroup.	77% (24/31) correct responses

Note: No student had more than two T/F items marked incorrect.

Table 6 continued

Question	Performance
ELEMENTS OF A GROUP	
Interview Item	
Given a commutative group G with an element of order 2 and an element of order 3, must G have an element of order 6? What would happen if 2 and 3 were replaced by other numbers?	46% (11/24) responded correctly. On a similar question asking whether all positive powers of a single element form a subgroup, 90% (28/31) responded correctly.
COSETS, NORMAL GROUPS, AND QUOTIENT GROUPS	
Item on Group Test	
State and prove Lagrange's theorem.	7/7 groups stated the theorem. With partial credit, the average score for the proof was 76%.
Items on Individual Test	
There are many conditions that are equivalent to a subgroup H of a group G being normal. One is: for all $g \in G$, $gHg^{-1} \subset H$. Give another condition of normality, and show that it is equivalent to this statement.	52% (16/31) supplied equivalent definitions.
Let $f : G_1 \longrightarrow G_2$ be a homomorphism. Prove or give a counterexample to the following statement: The kernel of f is a normal subgroup of G_1.	53% (16/30) provided a correct proof.
What does it mean for a subgroup of a group to be normal?	84% (26/31) understood that normality is a property of a subgroup.
Let S_3 be the group of permutations of three objects. Find a normal subgroup N of S_3, and identify the quotient group S_3/N.	84% (26/31) identified a normal subgroup of S_3. 77% (24/31) computed cosets of S_3/A_3 correctly. 73% (22/31) listed all of the cosets. 30% (9/30) identified the quotient group as $\mathcal{Z}_2$.
Let $\mathcal{Z}$ be the ring of integers. Describe the ring $2\mathcal{Z}/6\mathcal{Z}$.	90% (28/31) worked correctly with the two binary operations in the set of cosets.
Prove or disprove that the ring $2\mathcal{Z}/6\mathcal{Z}$ is isomorphic to $(\mathcal{Z}_3, +, \cdot)$.	90% (28/31) listed the cosets, mentioned or used the operation, and computed the Cayley table correctly.

Table 6 continued

Question	Performance
COSETS, NORMAL GROUPS, AND QUOTIENT GROUPS	
Interview Items	
In the group S_4, consider the subgroup $K = \{(1), (12)(34), (14)(23), (13)(24)\}$. How would you go about deciding if K is normal?	65% (15/24) applied the definition of normality to discuss what it would mean for K to be a normal subgroup.
Compute the operation table for S_4/K, $K = \{(1), (12)(34), (14)(23), (13)(24)\}$.	63% (15/24) computed sets of cosets, produced an operation table, and calculated coset products.
With respect to working with cosets, what gives you the right to use the more convenient method of representatives?	71% (17/24) provided an acceptable definition of the method of representatives for cosets.
PERMUTATION AND SYMMETRY GROUPS	
Item on Group Test	
Let D_4 be the group of symmetries of a square. Write out the elements of D_4.	7/7 groups received full marks.
Item on Individual Test	
What does it mean for a subset of S_n to be closed under the cycle structure?	61% (19/31) answered correctly. These 19 students, along with 10 others (94%), gave evidence of understanding what it would mean for two permutations to have a similar cycle structure.

The group test consisted of 7 items. Every group received full marks on 5 of the 7 items given on the group test. Of the two items in which full marks were nor awarded, the three groups that were not able to find all of the subgroups of D_4 did manage to find all but one or two of the subgroups. The other item, the first under Groups and Binary Operations, was originally part of a study of proof-writing performance reported by Hart (1994). Hart rated this particular question as the second most difficult of the six proofs of standard elementary group theory that were part of his study. Although only 3 of the 7 groups received full marks, the average group score, with partial credit awarded, was 89%.

Of the individual test items considered in Asiala et al. (1998), Asiala et al. (1997), and Brown et al. (1997), 77% of the students computed cosets of S_3/A_3 correctly; 73% found all of the cosets of S_3/A_3; 63% computed sets of cosets, formed an operation table, and calculated coset products for a quotient group of S_4; 90% worked correctly with the two binary operations in $2\mathcal{Z}/6\mathcal{Z}$; 94% gave evidence of understanding what it would mean for two permutation groups to have similar cyclic structure; and all of the students demonstrated evidence of having coordinated the two ring operations.

In all three algebra studies, the authors were interested primarily in learning how students might construct their understandings of particular topics. Asiala et al. (1998) considered permutations of a finite set and symmetries of a regular polygon. As a means of measuring students' understandings, the authors analyzed 24 students' responses to individual test and interview items using APOS Theory. In the case of permutations and symmetries, this meant determining whether each student was operating at an action, process, or object level. The percentage and the number of students working at each of the three levels is given in Table 7.

Table 7. Cognitive Levels: Permutations and Symmetries

Level	EXP Only	
	%	N
Object	42	10
Process	38	9
Action	21	5
Total interviewed		24

Each category specifies the limit of a given student's conception. For example, in reporting that 38% of the students possessed a process conception, the authors' analysis suggests that 9 of the 24 students interviewed were *limited* to a process conception.

Asiala et al. (1997) conducted a similar analysis of students' conceptualization of cosets. They determined each student's conception of cosets by considering her or his responses to the following three items, one from an individual test and the other two from the non-comparative interview. In this case, as well as the case just considered from Asiala et al. (1998), the authors determined each student's level, be it action, process, or object, through discussion and negotiation until consensus was achieved.

Table 8. Cognitive Levels of EXP Students: Cosets

	%	N
Test item: There are many conditions that are equivalent to a subgroup H of G being normal. One is: for all $g \in G$, $gHg^{-1} \subset H$. Give another condition, and show that it is equivalent to this statement.		
Results		
Object level	58	18
Process level	10	3
Action level	16	5
Little evidence of understanding	16	5
Total tested		31
Interview items: In the group S_4, consider the subgroup $K = \{(1), (12)(34), (14)(32), (13)(24)\}$. How would you go about deciding if K is normal? Compute the operation table for S_4/K.		
Results		
Object level	71	17
Process level	17	4
Between action/process level	13	3
Total interviewed		24

Permutation groups, cosets, and coset groups are arguably the most difficult concepts in introductory group theory. The results reported in this section suggest that instruction based upon APOS Theory may be effective in helping students to grasp the most challenging elements of a first course in abstract algebra.

3.3. Comparative Abstract Algebra Studies. At the beginning of Section 4, we mentioned that comparative data were used in the analyses comprising Asiala et al. (1997) and Brown et al. (1997). These data were gathered from individual, audio-taped interview sessions conducted with 17 students from the experimental group and all 20 students included in the traditional group. This group of 37 students served as subjects for both studies.

3.3.1. *Comparative Groups and Subgroups Studies.* Brown et al. (1997) compared the performance of the experimental and the traditional students on the two interview questions given in Table 9. Students' responses to each of these questions were analyzed by identifying indicators of successful and unsuccessful performance. Table 9 gives each indicator and the corresponding level of student performance.

Table 9. Performance on Center of a Group and Subgroups of $\mathcal{Z}$

	EXP		TRAD	
	%	$N = 17$	%	$N = 20$
Center of group interview item The student was given the definition of the center of a group (expressed in words), asked to express the definition in set notation, and then instructed to show that the center of a group is a subgroup. (Students were not required to say that the associative axiom was inherited; they could prove it directly. However, all who were successful on this item did mention that associativity was inherited.)				
Success indicators				
Constructed the center as a subset.	47	8	40	8
Understood how to apply group axioms to the center.	88	15	55	11
Realized that all 4 group axioms need to be checked.	76	13	5	1
Verified that center satisfies each group axiom.	35	6	15	3
Lack of success indicators				
Little progress in constructing the center as a subset.	35	6	30	6
Did not appear to know a single group axiom.	0	0	30	6
Unable to verify that the center satisfies group axioms.	18	3	20	4
Subgroups of $\mathcal{Z}$ interview item The student was asked to provide examples of subgroups of $\mathcal{Z}$, then was asked for a general statement. Once the characterization of subgroups of $\mathcal{Z}$ was on the table, the student was asked for a proof.				
Success indicators				
Knew every subset of the form $n\mathcal{Z}$ is a subgroup of $\mathcal{Z}$.	82	14	50	10
Knew all subgroups of $\mathcal{Z}$ have form $n\mathcal{Z}$.	47	8	20	4
Lack of success indicators				
Essentially no progress finding subgroups of $\mathcal{Z}$.	0	0	15	3
Thought that some $\mathcal{Z}_n$ groups are subgroups of $\mathcal{Z}$.	29	5	60	12

3.3.2. *Comparative Normality and Cosets Studies.* Asiala et al. (1997) obtained comparative results similar to those reported in Section 4.3.1 on a question involving normality. In this case, the authors specified five levels and assigned each student to a given level on the basis of her or his response to the following interview task.

Table 10. Cognitive Levels: Normality

		EXP		TRAD	
		%	N	%	N
Interview item:	Define the center of a group. Show that the center of a group forms a normal subgroup.				
Highest Level	The student provided a correct definition of normality and acknowledged the universal quantification condition.	29	5	5	1
	The student was able to state a condition for normality such as $gH = Hg$, with a tendency to move away from a static conception of the element g.	35	6	5	1
↓ ↓ ↓	The student was able to state a condition, such as $ga = ag$ or $gH = Hg$. In the former case, the student may have confused commutativity with normality. In the latter case, the student may have considered g to be a fixed element of G.	12	2	15	3
	The student thought of normality as an action applied to a subgroup, but the action conception was weak.	24	4	35	7
Lowest Level	The student demonstrated little or no understanding of the concept.	0	0	40	8
	Total Interviewed		17		20

In addition to normality, cosets were considered in the context of an interview item involving Lagrange's Theorem. The authors devised four indicators that measured student performance on this item. The specific interview item and comparative results are given in Table 11.

Table 11. Performance on Lagrange's Theorem Item

	EXP		TRAD	
	%	$N = 17$	%	$N = 20$
Interview item: 1. What do you remember about Lagrange's theorem? 2. What is the statement of Lagrange's theorem? 3. How does one prove Lagrange's theorem?				
Indicator				
States theorem.	41	7	20	4
Proved theorem.	35	6	15	3
Recalled an application of theorem.	65	11	20	4
Recalled or reconstructed conception of cosets without prompting.	53	9	15	3

On this question on Lagrange's theorem, 11/17 (65%) students from the experimental group and the 7/20 (35%) students from the traditional group discussed cosets in a meaningful way. The authors classified these students' cognitive levels in one of three ways.

Table 12. Cognitive Levels : Lagrange's Theorem

		EXP		TRAD	
		%	*N*	%	*N*
Highest Level	Demonstrated a fairly strong object conception.	45	5	43	3
↓ ↓					
	Demonstrated ability to reconstruct knowledge about cosets and to de-encapsulate; performance did not suggest a strong object conception.	27	3	14	1
↓ ↓					
Lowest Level	Discussed cosets as objects, but demonstrated no ability to de-encapsulate, or to apply actions to cosets.	27	3	43	3
	Totals		11		7

As with the calculus performance results, the results presented here favor students who completed a course based upon APOS Theory. Of the above 15 indicators of successful and unsuccessful performance used to measure students' understandings of cosets, groups, and subgroups, 14 of the 15 indicators favored the experimental group. The two measures indicating cognitive levels for cosets and normality also favored the experimental group. This is also consistent with additional non-comparative results on cognitive levels of development, which showed that 42% of the experimental students possessed an object conception of permutation groups, while nearly 60% gave evidence of an object conception of cosets.

4. Performance Results: The Concept of Function

Breidenbach et al. (1992) explored students' conceptions of function. Sixty-two sophomore and junior-level mathematics majors participated in this non-comparative study. These students, most of whom were pre-service secondary teachers, completed two discrete mathematics courses whose pedagogy was based upon APOS Theory.

At the beginning of each course, students were given a "pre-test" in which they were asked the question "What is a function?," followed by a request to give examples of different kinds of functions. The analysis focused upon determining whether each student was at an action, process, or object level. With respect to the general question "What is a function?," only 14% of the students provided process-level responses. Similarly, out of 106 student-generated examples, only 7.5% revealed a process-level conception. None of the students displayed object conceptions.

Students were asked the same two questions several weeks into the course. During the intervening period, they were assigned a large number of general computer activities involving computer-defined functions that were considerably different from standard algebraic/trigonometric expressions. The second administration occurred before any discussion of functions had taken place. It yielded the following results: with respect to the question "What is a function?," 36% indicated a process conception, and, of 177 student-generated examples of functions, 12% revealed a process-level conception.

Once the instructional treatment for functions was completed, the authors used two instruments to gather additional data. The first was an untimed, individual, written examination administered on a voluntary basis. Those students who volunteered to take the examination were informed that their performance would not affect their final course grade. The examination required students to construct a function and to use its underlying process to organize information, to coordinate complicated combinations of processes, and to reverse those processes.

The remaining post-instructional treatment data were student responses on the final course examination. The students were asked to explain the nature of a function, to provide examples of functions, to determine those functions that describe various situations, to reason about functions, to apply functions to various contexts, to solve problems involving sets of functions, and to perform operations on functions. The average scores for both tests are presented in Table 13.

Table 13. Performance of EXP Students on Functions

Question description	Average % score
Construction of Function to Organize Information	86
Complicated Function Expressed in Mathematical Programming Language (Non-numeric output)	
Interpretation of $F(a)(b)(c)$: F is a function whose value at a point a in its domain is also a function, denoted $F(a)$. $F(a)$, evaluated at a point b in its domain, is a function represented by $F(a)(b)$, whose value at a point c in its domain is the object $F(a)(b)(c)$. The final output d is non-numeric.	
• Evaluate $F(a)(b)(c)$, with values of a, b, c given.	73
• Given $F(a)(b)(c) = d$, find c when a, b, d given.	66
• Given $F(a)(b)(c) = d$, find b when a, c, d given.	75
• Given $F(a)(b)(c) = d$, find a when b, c, d given.	69
Composition of Functions $h = f \circ g$	
• Given information about f and g, evaluate $h(a)$ for some a.	84
• Given information about h and g, evaluate $f(a)$ for some a.	40
• Given information about h and f, evaluate $g(a)$ for some a.	87
Compute Product of Two Piece-Wise Functions	87
One-to-one/Onto (Prove or give counter-example)	
• If f, g are both 1-1, then $f + g$ is 1-1.	55
• If f, g are both onto, then $f + g$ is onto.	48
Operation/Actions Defined on Functions	
• Let $\Gamma = \{f : \mathbf{R} \longrightarrow \mathbf{R}\}$. Define D such that $\mathrm{D}(g) = g'$ for $g \in \Gamma$. Define K such that $\mathrm{K}(f) = f(-2x)$ for $f \in \Gamma$, $x \in \mathbf{R}$. Are D and K functions?	76
• If $f \in \Gamma$ such that $f(x) = x^3$, what is $\mathrm{K}(f)(3)$?	74
• Describe the inverse of K.	31

Students could not answer questions in the last category, Operations/Actions Defined on Functions, without having first achieved an object conception of function. The results reported here suggest that at least 70% of the students progressed to a level where they could answer at least one of these items correctly, despite the fact that prior to instruction no student demonstrated an object conception, and only 14% gave evidence of a process-level conception.

5. Performance Results: Quantification

Dubinsky (1997) and Dubinsky et al. (n. d.) discussed students' understandings of quantification. In both cases, only non-comparative data were collected from students who completed courses using APOS Theory.

5.1. On Learning Quantification. Thirty-six students pursuing joint majors in mathematics and computer science participated in the study (Dubinsky, 1997). Of these students, 19 were enrolled in a sophomore-level course (Class 1) in which quantification was a major topic, and the remaining 17 students were enrolled in a junior-level course (Class 2) in which quantification was only treated briefly.

In the sophomore-level class (Class 1), data were students' responses to a combination of three in-class, closed-book, individual assignments and two open-book, take-home assignments. Among the five assignments, a total of 21 multiple-part questions on quantification were asked. In the junior-level course (Class 2), data were students' responses to a take-home assignment consisting of 16 questions. In both classes, students were asked to translate, from natural to formal language and vice-versa, statements using two or more levels of quantification; to negate statements including quantifiers; and to reason about statements containing quantifiers once various conditions were changed or introduced.

The average examination scores for both classes are given in Table 14. On each question, 100 points was awarded for a correct response, 50 points was given for a partially correct response, and the student received 0 points for an incorrect response. For each bulleted subtopic, the number of scores that are reported corresponds to the number of questions asked. Each score gives the per student class average for each question.

Table 14. Quantification Performance of EXP Students

Question description	Class 1	Class 2
Translating Quantification Statements		
• Convert statement from English to formal language.	69	76, 88, 94
• Negate formal language statement, express in English.	56, 64	
• Translate English statement into mathematical programming language, perform negation, and translate back into English.	71, 81, 89, 94	12, 29, 53
• Express English statement in mathematical programming language.	47, 84	
Negating Statements (no translation)		
• Negate English statement.	58, 95	41, 44
• Negate formal statement.	55	65
Reason about Statements Containing Quantifiers		
• Discussion of truth/falsity after altering statement.	56, 56, 58, 61, 82, 89, 100	15, 82, 85, 100
• Given formal language statement, explain conditions yielding falsity.	76	
• Given English statement, determine how computer or person would determine truth/falsity.	75	
• Discussion of equivalence of two statements		35, 47, 97

Of the 37 items above given to both classes, the average score exceeded 70 on 18 items. With a score of 50 given for a partially correct response, one could argue that the two groups of students made reasonable progress on nearly every task with the exception of translating English statements into a mathematical programming language.

5.2. Unpublished Study of AE and EA Quantification.
Dubinsky et al. (n. d.) studied 23 junior and senior mathematics majors who were taking a sophomore/junior level course in discrete mathematics that used various aspects of APOS Theory. Most of the students involved in the study were pre-service secondary mathematics teachers.

The non-comparative data in this study were student responses to a multi-part examination question that each student worked on individually. The question was based upon the following quantification statement.

> To each odd integer i between 10 and 1000 there corresponds an integer j in the set $\{3, 5, 7, 9\}$ such that if j divides i then i is less than 500.

The test items related to this statement were graded on a 10 point scale, with partial credit given. The average scores for these test items are reported in Table 15.

Table 15. AE and EA Quantification Performance of EXP Students

Question description of Task	Average Score
Express statement in formal language.	100
Express negation of statement in formal language.	100
Express negation of statement in English.	90
Is the original statement true or false? Explain.	60
What can you say about truth/falsity of the statement if "odd integer" is replaced by "prime number"?	90

With the exception of determining the truth or falsity of the original statement, the students in this study achieved a high degree of success on each task.

The results reported in the two studies considered in this section show that the students who received instruction on quantification using APOS Theory and delivered through the ACE Teaching Cycle made definite progress in their ability to work with statements involving quantifiers.

6. Student Performance: Mathematical Induction

Dubinsky (1989) investigated students' abilities to construct and to discuss mathematical induction proofs after having completed a sophomore-level discrete mathematics course that was taught using pedagogy based upon APOS Theory. Non-comparative data were collected from a total of 40 students who took such a course at two different universities. Most of the participants were pursuing majors in computer science or engineering.

Two instruments were used to collect data. The first was a take-home, individual exam consisting of 10 problems involving mathematical induction. The items comprising this exam are given below.

(1) Show that $(n^4 - n^2) \bmod 3 = 0$.

(2) Prove that $(1)(3) + (2)(4) + (3)(5) + \cdots + n(n+2) = \dfrac{n(n+1)(2n+7)}{6}$.

(3) Show that $\displaystyle\sum_{i=1}^{n} \frac{1}{i(i+1)} = \frac{n}{n+1}$.

(4) Consider the following ISETL procedure.

```
proc compute(x,y); z:=x; w:=y; (while w > 0) z:=z+y; w:=w-1;
end while; return z; end proc;
```

Show that the following loop is invariant for the beginning of the while loop:

$$yw + z = x + y^2.$$

(5) Show that in a casino with chips with \$5 and \$9, any sufficiently large amount of dollars can be represented.

(6) Show that an integer consisting of 3^n identical digits is divisible by 3^n.

(7) Show that the following inequality holds for n sufficiently large:

$$n - 2 < \frac{n^2 - n}{12}.$$

(8) Formulate and prove a general formula stemming from the following observations:

$$\begin{aligned} 1^3 &= 1 \\ 2^3 &= 3+5 \\ 3^3 &= 7+9+11 \\ 4^3 &= 13+15+17+19. \end{aligned}$$

(9) Let σ be an alphabet and A a language in σ^*, with the property that $A^2 = A$. Show that $A^* = A$.

(10) Show that $11^{n+2} + 12^{2n+1}$ is divisible by 133.

The students were given one week to complete the exam. They were asked to work entirely on their own without interaction with other students and without use of any source material other than their class notes and textbook. Of the 400 student proofs that were submitted, 213 were correct. With partial credit, the average individual score was 71%.

The second instrument used to collect data was the subsequent follow-up essay examination. Students were given a list of questions in which they were asked to describe the process and meaning of induction. Those questions are given below.

(1) Describe each of the following terms and give an example for each.
   a. Method of proof
   b. Proposition-valued function of the positive integers
   c. Implication-valued function of the positive integers

(2) Explain, in your own words, what it means to make a proof by induction. What is involved in making such a proof?

(3) Modus Ponens (method of the bridge) refers to the mental activity by which, if you know an implication $A \Rightarrow B$ and you know that A holds, then you know that B holds. What role does this concept play in Mathematical Induction?

(4) Suppose you have applied the method of Mathematical Induction to some statement, starting, say, with the first value equal to 1, and successfully completing the proof. Having done this, what would you say to someone who didn't know much about mathematics to convince her or him that, because of your proof, the statement is true for a fixed integer, say $n = 10$? What about $n = 1000$? Give the most concrete way that you can of trying to convince the person.

Of the 40 students involved in the study, 38 took the essay examination involving the four items given above. The author evaluated student responses using six criteria. Percentages of responses satisfying each criterion are presented in Table 16.

Table 16. EXP Student Responses to Essay Examination

Criterion	%	$N = 38$
Had an idea of the method of proof as a general tool.	74	28
Indicated understanding of function in which positive integers are input and boolean values are output.	74	28
Indicated understanding of function in which positive integers are input and implication is output.	76	29
Gave a reasonable explanation of proof by induction.	87	33
Provided a satisfactory explanation of the role of modus ponens.	71	27
Explained the operation of determining the truth of a statement statement for a single positive integer and how to apply modus ponens ad infinitum for all higher integer values.	58	22

Many students encounter considerable difficulty in learning how to write induction proofs. Even those who are able to write such proofs often do not understand how the initializing step and the conditional statement linking the kth and $(k+1)$st steps work together to establish the truth of a proposition-valued statement for all positive integers beyond the initializing step. The analysis of responses to the first take-home test suggest that instruction using APOS Theory helped students to achieve success in writing induction proofs. The results of the essay test revealed that a sizable proportion of the students having received this instruction understood the components of a proposition-valued function and the role of modus ponens, as well as the ability to explain how to apply modus ponens ad infinitum.

7. Student Attitudes and Long Term Effects

It is important to consider attitudes and performance together. It is not very difficult to conduct a course that students feel good about. The real goal is for students to feel good about developing a profound understanding of non-trivial mathematics concepts. This issue cuts across all of the individual mathematics courses we have considered in this paper.

In this section we present the results of three studies of the attitudes students developed as a result of having taken courses based on the approach described in this paper. We also will consider the long-term effects of such instruction in terms of both attitudes and performance. All of the statistics presented here are comparative, with data gathered from students who completed courses based upon APOS Theory (EXP) contrasted with data from students having completed courses using traditional curricula and pedagogies (TRAD).

In Section 8.1, we compare the attitudes of students who completed an APOS/ACE abstract algebra course with the attitudes of students who took a traditional course in abstract algebra. In Section 8.2, we consider certain aspects of students' mathematical performances over a six-year period by comparing the performance of students who completed an APOS/ACE course in calculus with the performance of students who completed a traditional calculus course. Finally, in Section 8.3, we present comparative analysis of attitudes of students who took courses in precalculus, calculus, discrete mathematics, and abstract algebra several years after the students completed these courses. These courses used some or all of the components of the APOS/ACE approach. Although specific performance data were not collected for this group, a cursory review of the evaluative instruments used in these courses suggests that student learning in these contexts did not differ in any significant degree with that of the experimental student groups considered in the earlier sections of this paper.

7.1. Student Attitudes in Abstract Algebra. This section describes a study of the effect of APOS-based instruction in abstract algebra on student attitudes. During the interview phase of the abstract algebra studies, the interviewers asked each student to describe her or his feelings about abstract algebra as a way of helping each student to feel at ease. Responses to this question, as well as other unsolicited remarks made by students during the interview process, yielded important information.

Although the information presented below will show that the experimental students were more positive about mathematics than the traditional students, the data may be skewed in favor of experimental results in the following respects:

- Both experimental abstract algebra courses were taught by the same instructor (one of the developers of the method), whereas the traditional courses were taught by three different instructors about whom little information was obtained.
- The investigation of attitudes was not systematic; no attempt was made to perform reliability tests.
- The interviews were conducted after completion of the course by the instructor of the experimental course or a graduate student who had directed laboratory sections of the experimental course.
- The grades in the experimental courses were higher overall than those in the traditional courses (GPA of 3.1 out of 4 vs. 2.3 out of 4).

On the other hand, because the APOS/ACE pedagogical approach emphasizes conceptual development within a student-centered, experimental, active learning context, students are expected to be more self-reliant academically. As a result, there was a possibility that some students, particularly those who had never experienced such reform-based mathematics courses, might have been more likely to complain about a heavier work load and a lack of teacher-centered instruction. However, the results, as presented in Table 17, do not indicate this.

Table 17. Student Attitudes in Abstract Algebra

Interest indicator	EXP		TRAD	
	%	*N*	%	*N*
Positive				
Had immediate positive reactions to the course.	73	8	32	7
Made positive comments pertaining to computer work.	64	7	N/A[a]	
Believed he/she learned a great deal.	73	8	0	0
Negative				
Made exclusively negative comments about course.	0	0	18	4[b]
Considered course difficult.	27	3	64	14
Total interviewed		11		22

[a]No computer technology was used in the traditional course.
[b]These students received a grade of B or better for their course.

In addition to the information given in Table 17, several other observations were made.

- Two students took both the traditional course and the experimental course. Both students took the experimental course after having failed the traditional course. These students claimed that they learned much more from the experimental course.
- 5/22 (23%) students from the traditional courses and 4/11 (36%) from the experimental course commented about course grades or tests. Test performance seemed to be more on the minds of the traditional students; their comments centered on passing tests rather than with what they had learned. Experimental students concentrated on what they felt they had learned even when their test performances and course grades were low.
- In some cases, students from the traditional courses displayed emotionally intense negative attitudes that were not observed in students from the experimental course.
- A majority of the traditional students made negative comments about the course.
- The majority of comments made by the experimental course students were positive, with a focus upon how much they had learned.

7.2. Long-Term Effects in Calculus. Educational innovation can sometimes be problematic because the improvements that are observed in student learning and performance may only be temporary. After the Hawthorne effect has diminished and the students return to traditional pedagogy, they may not only fail to continue functioning at a higher intellectual level, but all evidence of the experimental improvements may disappear (for an explanation of the Hawthorne effect, see Krathwohl, 1993). Furthermore, the additional effort expended in an experimental course by both the instructor and the students may not transfer positively to later, more traditional courses. Kuhn et al. (2000) attempted to investigate these issues by conducting a statistical study of various aspects of student performance. The study involved comparing the performance of students who received introductory calculus instruction using C^{4}L (EXP) materials to that of students who received instruction via a traditional lecture/recitation method (TRAD). All of the students involved in the study were enrolled at a major midwestern research university. Researchers reviewed the records of 4636 students, 205 of whom participated in the experimental group and the remainder of whom took traditional courses. The calculus courses in question covered the period from 1988 to 1991.

The specific tasks of the project involved comparing course grades, the extent to which students continued their studies in calculus, and the number of mathematics courses that students pursued beyond calculus. In making these comparisons, the researchers used an additive multivariable multiple regression model that allowed them to control for possible confounding variables such as gender, major course of study, and predicted grade point average. The statistically "adjusted" results for each of the categories considered is provided in Table 18. The middle column presents the 95% confidence interval for the difference in means between the experimental group and the traditional group. For instance, the first confidence interval of $.42 \pm .06$ expresses the idea that the average final grade (based on a 4.0 scale) of all EXP students was between .36 and .48 grade points higher than the average final grade of all TRAD students. The final column gives the p-value associated with a hypothesis test for the difference in means, where the null hypothesis represents no difference.

Table 18. EXP and TRAD Students' Grades and Participation

Category	95% CI	p-value
Average final grade		
Calculus	0.42 ± 0.06	0.0001
All math courses after calc II	no statistically significant difference	
Last available overall grade point average	0.09 ± 0.05	0.05
Average number of courses taken		
Calculus	0.24 ± 0.05	0.0001
Mathematics other than calculus	0.16 ± 0.07	0.025

Based upon the statistics presented in Table 18, the following conclusions can be drawn:

- The EXP students earned higher grades in calculus courses; their grades were almost one-half letter grade higher, on average, than the TRAD students' grades.
- The EXP students progressed through more of the calculus sequence than did the TRAD students.
- The EXP students took slightly more non-calculus mathematics courses on average than did the TRAD students.

7.3. Long-Term Attitudes from Several Courses. During the period 1996–1999, a total of 11 courses in Pre-calculus (one section), Calculus I (three sections), Calculus 2 (three sections), Discrete Mathematics (two sections), and Abstract Algebra (two sections) were taught using various components of the experimental approach to students attending a large, urban, comprehensive university in the southeastern United States. At the same time, one or more sections of each of these courses was taught by regular faculty members, part-time instructors, and graduate teaching assistants using a traditional approach. The experimental sections were taught by five individuals: two faculty members (one of whom is one of the authors of this article), one part-time instructor, and two graduate teaching assistants.

7.3.1. *Population.* The results of the study are presented here for the first time. Data were gathered and analyzed by Mark Asiala, who was one of the graduate students teaching the EXP sections (Asiala & Dubinsky, 1999). The total population, about 1800 individuals from 75 sections, consisted of those students who completed the courses in previous semesters or who were two-thirds through the course in the semester in which the data were gathered. Generally speaking, these students were rather weak mathematics students, even in those cases in which their major was mathematics.

Since it was not feasible to send out a questionnaire to the entire population, a random sample of 400 individuals was selected and stratified by section so that it was possible for the sample to be representative of all of the sections, including the ratio of APOS/ACE sections to traditional sections. Questionnaires were mailed to the students in this sample and they were instructed to return them within 10 days in the postage-paid envelopes provided. A total of 90 responses was received, a 23% response rate. The responses broke down nearly 50/50 between the APOS/ACE group and the traditional group.

7.3.2. *Questionnaire.* Each of the questions given Table 19 was evaluated using the Likert scale (Likert, 1940). Many of the questions have been used in other surveys. The researchers who conducted this study did not perform their own validity tests on the questions, but relied generally on the body of research which produced them. The results of the questionnaire, together with each of the questions, are presented in Table 19.

Table 19. Questionnaire Items and Responses

	Question	APOS/ACE		TRAD		Desired	
		Mean	Var	Mean	Var	outcome	p-value
1.	I enjoy doing math problems.	4.09	1.06	3.72	0.92	+	0.039
2.	To solve math problems you have to know the exact procedure for each problem.	3.07	1.51	3.17	1.35	–	0.338
3.	In mathematics, an answer is either right or wrong.	3.07	1.73	3.54	1.36	–	0.038
4.	Mathematics requires much more memorization than understanding.	1.70	0.68	2.15	1.02	–	0.012
5.	In the long run, I think mathematics will help me.	4.39	0.75	4.28	0.56	+	0.273
6.	In order to understand math, I need to know more theory.	3.33	1.18	3.24	1.12	+	0.352
7.	I learn math through examples.	4.55	0.44	4.51	0.26	–	N/A
8.	Guessing is OK in solving mathematics problems.	2.67	1.27	2.35	0.90	+	0.071
9.	My calculus course seems to require more thinking than memorization.	3.86	0.88	3.39	1.18	+	0.016
10.	I think that I can apply what I'm learning in calculus in some other courses.	3.76	1.09	3.42	1.16	+	0.065

Table 19 continued

	Question	APOS/ACE		TRAD		Desired	
		Mean	Var	Mean	Var	outcome	p-value
11.	The best way to learn calculus is to memorize all the formulas.	2.20	0.86	2.57	1.01	–	0.040
12.	Some people are good in calculus, some just aren't.	2.77	1.30	3.17	1.08	–	0.042
13.	I like the idea of using technology in calculus.	3.86	1.19	3.59	0.96	+	0.104
14.	The math teacher is responsible for how much math I learn.	3.18	1.50	2.83	1.39	–	N/A
15.	A good math teacher demonstrates procedures for doing the type of problems that'll be on the test.	3.93	1.23	4.13	0.74	–	0.173
16.	Good math teachers show students lots of different ways to look at the same question.	4.55	0.30	4.14	0.67	+	0.004
17.	I like the idea of working in groups in a calculus course.	3.57	1.79	3.22	1.69	+	0.105
18.	I feel uncomfortable when I see that other students know more math than I do.	2.68	1.66	3.00	1.24	–	0.107

7.3.3. *Analysis of Questionnaire Data.* Given a sample size of approximately 45 for each of the two groups of responses, it is reasonable to assume that the responses will have an approximate normal distribution. Thus, in comparing the two groups, a one-tail t-test was used to compare the sample means. The table gives the means, variances, and p-values for the one-tail t-test. It also indicates, via a $+$ or a $-$ in the Desired Outcome column, whether it is better to have a high score or a low score. Thus, for example, in Question 1, we would prefer a high score, whereas in Question 2, a low score indicates a better attitude towards problem solving.

We can summarize these results by noting that with the exception of Questions 7 and 14, the result, in every case, whether high or low, favors the APOS/ACE classes. In some cases (Questions 1, 3, 4, 9, 11, 12, 16), the differences were statistically significant at the $p = 0.05$ level of significance. Because of technical difficulties, tests of significance were not made for Questions 7 and 14.

8. Concluding Remarks and Observations

Several factors potentially affect the reliability of the results presented in this report: lack of triangulation; small sample sizes for quantitative data; lack of randomization in student selection; potential lack of objectivity on the part of the interviewers; lack of control in specific subject matter coverage between the experimental and the traditional groups; and potential bias in the formulation of interview questions in favor of the experimental groups. Nonetheless, the results are striking, consistent, and broad-based in favor of the experimental groups. In particular:

- The comparative calculus and abstract algebra studies consistently point in the direction of the experimental groups. In essentially every category under consideration, students from the experimental groups achieved a

higher level of performance on fairly standard tasks than did students from the traditional groups.
- With respect to the non-comparative statistics, students who received instruction based upon APOS Theory and the ACE Teaching Cycle performed at a level consistent with what was reported in the comparative analysis. Based upon our collective teaching experience, we believe the non-comparative results reported in this paper are higher than those one would typically expect from students having completed similar traditionally-structured courses.
- In terms of cognitive development, students from the experimental groups appeared to have achieved conceptions as advanced, and often more advanced, than students from the traditional groups.

In addition, instruction based upon APOS Theory seems to generate student interest and enthusiasm for mathematics, as well as potential, sustained long-term academic benefits. Students who completed the experimental abstract algebra courses tended to focus attention upon their learning, as opposed to utilitarian aspects of coursework, such as performance on tests. There does not appear to have been a significant Hawthorne effect. Students who received APOS-based instruction in calculus performed at least as well in subsequent mathematics courses as students who received traditional calculus instruction. In addition, students who completed APOS calculus courses were more likely to pursue additional study of mathematics.

Given the wide range of topics covered in these papers and the consistency of the performance results, one could conclude that the direction of the results in favor of the experimental groups is probably not due to chance alone. These results, along with the non-comparative results, suggest that instruction based upon APOS Theory yields better results than what one would expect within a traditional setting. Moreover, these papers provide strong evidence supporting the contention that the research and curriculum development framework based on APOS Theory and the ACE Teaching Cycle is a reasonable approach to describe and to enhance student learning of mathematics.

Our results seem to point to the success of this theoretically-based approach as an effective tool for helping students learn advanced mathematical concepts. However, the results reported in this study were culled from a number of research papers whose primary methods were qualitative and whose primary aim was to reveal the nature of students' understandings rather than to compare students' performances statistically. A statistically sound, quantitative, independent comparison study could build upon these results. The authors call for other researchers in collegiate mathematics education to conduct further comparative evaluations of the performance of students who have completed APOS-based courses with those who have received traditionally-based instruction. Further, it is hoped that the results presented in this paper will motivate teachers of mathematics to learn more about these approaches and to incorporate these ideas into their classes.

Finally, it may be asked why our approach can be successful. Our conjecture is that APOS Theory provides one way to describe the process of learning various mathematical concepts. Moreover, the ACE Cycle is designed to fit exactly with the genetic decompositions resulting from an APOS analysis. That is, we suppose that writing certain kinds of computer programs provides direct stimulus for the interiorizations and encapsulations called for by an APOS analysis. In addition, co-operative learning fosters reflection, which, according to APOS Theory, is a vitally important aspect of a student's ability to make mental constructions and to form connections between related concepts. Thus, our hypothesis is that instructional treatment based on APOS Theory and delivered using the ACE Teaching Cycle *should* lead to improvements in student learning.[2] The results presented in this paper are consistent with that hypothesis.

References

Asiala, M., Brown, A., DeVries, D., Dubinsky, E., Mathews, D., & Thomas, K. (1996). A framework for research and curriculum development in undergraduate mathematics education. In J. Kaput, A. H. Schoenfeld, & E. Dubinsky (Eds.), *Research in collegiate mathematics education II* (pp. 1–32). Providence: American Mathematical Society.

Asiala, M., Brown, A., Kleiman, J., & Mathews, D. (1998). The development of students' understanding of permutations and symmetries. *International Journal of Mathematical Learning, 3*, 13–43.

Asiala, M., Cottrill, J., Dubinsky, E., & Schwingendorf, K. (1997). The development of students' graphical understanding of the derivative. *Journal of Mathematical Behavior, 16*(4), 399–431.

Asiala, M., & Dubinsky, E. (1999). *Evaluation of research based on innovative pedagogy used in several mathematics courses.* (Unpublished report, available from the authors)

Asiala, M., Dubinsky, E., Mathews, D., Morics, S., & Oktac, A. (1997). Student understanding of cosets, normality, and quotient groups. *Journal of Mathematical Behavior, 16*(3), 241–309.

Baker, B., Cooley, L., & Trigueros, M. (2000). A calculus graphing schema. *Journal for Research in Mathematics Education, 31*(5), 557–578.

Beth, E. W., & Piaget, J. (1966). *Mathematical epistemology and psychology.* Dordrecht: W. Mays and Reidel. (Original work published 1965; English translation)

Breidenbach, D., Dubinsky, E., Hawks, J., & Nichols, D. (1992). Development of the process conception of function. *Educational Studies in Mathematics, 23*, 247–285.

Brown, A., DeVries, D., Dubinsky, E., & Thomas, K. (1997). Learning binary operations, groups, and subgroups. *Journal of Mathematical Behavior, 16*(3), 187–239.

Clark, J., Cordero, F., Cottrill, J., Czarnocha, B., DeVries, D., John, D. S., Tolias, G., & Vidaković, D. (1997). Constructing a schema: The case of the chain rule. *Journal of Mathematical Behavior, 16*(4), 345–364.

Clark, J., DeVries, D., Litman, G., Meletiou, M., Morics, S., Schwingendorf, K., & Vidaković, D. (2000). *A story of progress: Research in undergraduate mathematics education.* (Submitted for publication)

Clark, J., Hemenway, C., John, D. S., Tolias, G., & Vakil, R. (1999). Student attitudes toward abstract algebra. *PRIMUS, 9*(1), 76–96.

---

[2]Readers interested in learning more about such instructional treatments should consult Dubinsky and Leron (1994), Dubinsky and Schwingendorf (1996), Dubinsky et al. (1995), Fenton and Dubinsky (1996), Reynolds et al. (1996), and Schwingendorf et al. (1996).

Dubinsky, E. (1989). On teaching mathematical induction II. *Journal of Mathematical Behavior*, *8*, 285-304.

Dubinsky, E. (1995). ISETL: A programming language for mathematics. *Communications on Pure and Applied Mathematics*, *48*, 1027–1051.

Dubinsky, E. (1997). On learning quantification. *Journal of Computers in Mathematics and Science Teaching*, *16*(2/3), 335–362.

Dubinsky, E., & Leron, U. (1994). *Learning abstract algebra with ISETL.* New York: Springer-Verlag.

Dubinsky, E., Litman, G., Morics, S., & Oktac, A. (n. d.). *Data from unpublished study on students' understanding of quantification.* (Study unfinished)

Dubinsky, E., & Schwingendorf, K. (1991). Constructing calculus concepts: Cooperation in a computer laboratory. In C. Leinbach (Ed.), *The laboratory approach to teaching calculus* (MAA Notes no. 20, pp. 47–70). Washington, DC: Mathematical Association of America.

Dubinsky, E., & Schwingendorf, K. (1996). *Calculus, concepts, and computers: Multivariable and vector calculus* (Preliminary Version). Raleigh: McGraw-Hill.

Dubinsky, E., Schwingendorf, K., & Mathews, D. (1995). *Calculus, concepts & computers.* New York: McGraw-Hill.

Fenton, W., & Dubinsky, E. (1996). *Introduction to discrete mathematics with ISETL.* New York.

Hart, E. (1994). Analysis of the proof-writing performances of expert and novice students in elementary group theory. In E. Dubinsky & J. Kaput (Eds.), *Research issues in mathematics learning: Preliminary analyses and results* (MAA Notes no. 33, pp. 49–62). Washington, DC: Mathematical Association of America.

Krathwohl, D. R. (1993). *Methods of educational and social science research: An integrated approach.* White Plains: Longman Publishers.

Kuhn, J., McCabe, G., & Schwingendorf, K. (2000). A longitudinal study of the C^4L reform program: Comparisons of C^4L and traditional students. In E. Dubinsky, A. H. Schoenfeld, & J. Kaput (Eds.), *Research in collegiate mathematics education IV* (pp. 63–76). Providence: American Mathematical Society.

Likert, R. (1940). A technique for measurement of attitudes. *Archives of Psychology*, *140*, 1–55.

McDonald, M., Mathews, D., & Strobel, K. (2000). Understanding sequences: A tale of two objects. In E. Dubinsky, A. H. Schoenfeld, & J. Kaput (Eds.), *Research in collegiate mathematics education IV* (pp. 77–102). Providence: American Mathematical Society.

Piaget, J., & Garcia, R. (1989). *Psychogenesis and the history of science.* New York: Columbia University Press. (Original work published 1983; English translation by H. Feider)

Reynolds, B., Przybylski, J., Kiaie, C., Schwingendorf, K., & E.Dubinsky. (1996). *Precalculus, concepts & computers.* New York: McGraw-Hill.

Schwingendorf, K., Mathews, D., & Dubinsky, E. (1996). *Applied calculus, concepts, and computers* (Preliminary Version). Raleigh: McGraw-Hill.

Department of Mathematics, University of North Texas, Denton, TX 76203
E-mail: wellerk@unt.edu

Department of Mathematics and Statistics, Hollins University, Roanoke, Virginia 24020
E-mail: jclark@hollins.edu

265 North Woods Rd., Hermon, NY 13652
E-mail: edd@mcs.kent.edu

Department of Mathematics and Computer Sciences, Grand View College, Des Moines, IA 50316
E-mail: sloch@gvc.edu

Department of Mathematics, Occidental College, Los Angeles, CA 90041
E-mail: mickey@oxy.edu

Department of Mathematics, Computer Science, and Statistics, Purdue University Calumet, Hammond, IN 46323
E-mail: merk@calumet.purdue.edu

CBMS Issues in Mathematics Education
Volume **12**, 2003

Models and Theories of Mathematical Understanding: Comparing Pirie and Kieren's Model of the Growth of Mathematical Understanding and APOS Theory

David E. Meel

ABSTRACT. The search for a meaningful cognitive description of understanding has ensued for the past half a century. Within the past three decades, new and integrative perspectives have grown out of Richard Skemp's distinctions between instrumental and relational understanding. The growth of these perspectives, up until 1987, was documented by Tom Schroeder in his PME synthesis of the work on understanding resulting from Richard Skemp's instrumental/relational contrasts. Since 1987, the work on understanding has progressed and this paper examines the new, more recent theoretical frameworks of understanding which have arisen from these roots. This paper focuses on two theoretical frameworks, Pirie and Kieren's model of the growth of mathematical understanding and Dubinsky's APOS theory, and discusses other contemporary theoretical frameworks such as the work by Cornu and Sierpinska on cognitive or epistemological obstacles, the investigations into concept definition and concept image by Vinner and Tall, Kaput's explorations of multiple representations, and Sfard's distinctions between operational and structural conceptions. Besides explicating the definitions of understanding proposed by these two frameworks, the discussion addresses their elements and constructs as well as their linkages to historical and recent characterizations of understanding. The paper then argues why Pirie and Kieren's model and APOS theory satisfy the Schoenfeld (2000) criteria for classification as a theory and finally concludes with discussions of a variety of interconnections between these two theories as well as the elements which make them distinct from each other such as their origins, organizations, relationships to other frameworks, and implications of the two theories for both assessment and pedagogical practices.

This paper opens with a brief history of the search to obtain a clearly defined conceptualization of the meaning of the term "understanding" consistent with the

---

As is the case with any review and synthesis, this paper can only draw upon the author's interpretations of published snapshots of these evolving theories. As a result, the interpretations and conclusions drawn herein address a temporal look at continuously changing theories by providing a state of the theory look to each theory's accumulated and self-supported published writings.

In addition, the author would like to thank the members of his Mathematics Education seminar that included Dr. Barbara Moses, Oxana Grinevitch, Becky Kessler, Daria Fillipova, Rachel Rader, and Julie Grabowski while extending special thanks to Ed Dubinsky, Susan Pirie, and the anonymous reviewers for their very helpful and insightful comments on earlier versions of this paper.

work of Schroeder (1987). Historically, various characterizations have associated understanding with knowledge, linkages built from mathematical operations, an act, or simply a grasping of meaning. The discussion contrasts the thinking prior to Skemp's influential work distinguishing the difference between instrumental and relational understanding with the various theoretical frameworks that have since developed. In particular, a brief description of the viewpoints taken by Brownell and Polya as well as their contemporaries provides a backdrop for examining the movement to distinguish between understanding and knowledge. This chain culminates in an abridged discussion of four recent theoretical frameworks addressing the development and acquisition of mathematical understanding thereby setting a stage for sections describing Pirie and Kieren's model of the growth of mathematical understanding and Dubinsky's APOS theory.

In brief, Pirie and Kieren (1991a) have aligned themselves with Schoenfeld's (1989) assessment of understanding as an unstable and retrogressive organic element and consider understanding as a whole, dynamic, layered but non-linear, never-ending process of growth. They reject the notion of the growth of understanding as a monotonically increasing function and consider it as a dynamic, organizing, and reorganizing process (Pirie & Kieren, 1992b, 1994b). On the other hand, Dubinsky (1991) aligns APOS theory with the Piagetian perspective that *reflective abstraction* is the key to cognitive development of logico-mathematical concepts.[1] From the perspective of APOS theory, understanding is a never-ending process of iterative schema construction through reflective abstraction, a cognitive process where physical or mental actions are reconstructed and reorganized by the learner on a higher plane of thought and thereby become understood by that learner (Ayers, Davis, Dubinsky, & Lewin, 1988). In addition to expounding these ideas, this discussion addresses the elements and constructs of the theories. With such components in mind, brief illustrations identify how their respective theories are related to both historical and recent characterizations of understanding.

The paper concludes with a discussion of a variety of interconnections between Pirie and Kieren's model of understanding and APOS theory. Both of these theories have constructivist origins but also contain elements that differentiate them. The focus then turns to the organizational structures of the two theories with particular attention paid to the elements and constructs of each the two theories. Next, the discussion expounds on the linkages to recent theoretical frameworks of understanding as well as a connection between Pirie and Kieren's model of understanding with a particular refinement of APOS theory defined by Clark et al. (1997). The final element of this section compares and contrasts the implications of the two theories for both assessment and pedagogical practices.

1. A brief history of "understanding"

Even though the term "understanding" has been freely used in mathematics education literature, the search for a concise definition of "understanding" has been going on for years. In particular, Brownell and Sims (1946) felt mathematical understanding was a difficult concept to define and stated, "A technically exact definition

[1]A logico-mathematical concept is one where the physical properties of objects have been abstracted and integrated into a learner's mental framework through physical experience. Analysis of this experience takes place through thought processes about the physical experience rather than manipulation of the physical objects from which the concept was derived.

of 'understand' or 'understanding' is not easily found or formulated" (p. 163). However, many writers have operated under the assumption that a well-defined notion of understanding existed, thereby causing entanglements with philosophy when they abandoned speaking of understanding in an ideal sense and attempted to explicate its meaning (Sierpinska, 1990b). These difficulties, according to Sierpinska (1990a), arose from the mathematics education community's inability to distinguish between knowledge and understanding prior to Skemp's (1976) famous paper on "Instrumental and relational understanding." In particular, it was only until a 1978 reprint in the *Arithmetic Teacher* that Skemp's distinction between knowledge and understanding drew the attention of the U. S. mathematics education community.

1.1. "Understanding" prior to 1978. Prior to Skemp's influential paper, U. S. researchers generally identified understanding with knowledge. Understanding became equated with the development of connections in the context of performing algorithmic operations and problem solving (Brownell, 1945; Brownell & Sims, 1946; Fehr, 1955; Polya, 1945; Van Engen, 1949; Wertheimer, 1959). For instance, Brownell and Sims (1946) characterized understanding as (a) an ability to act, feel, or think intelligently with respect to a situation; (b) varying with respect to degree of definiteness and completeness; (c) varying with respect to the problem situation presented; (d) requiring connections to real-world experiences and the inherent symbols; (e) requiring verbalizations although they may contain little meaning; (f) developing from varied experiences rather than repetitive; (g) influenced by the methods employed by the teacher; and (h) inferred from observations of actions and verbalizations. Polya (1962), on the other hand, identified understanding as complementary to problem solving as indicated in the following quotation:

> One should try to understand everything; isolated facts by collating them with related facts, the newly discovered through its connections with the already assimilated, the unfamiliar by analogy with the accustomed, special results through generalization, general results by means of suitable specialization, complex situations by dissecting them into their constituent parts, and details by comprehending them within a total picture. (p. 23)

This understanding, according to Polya (1962), cannot be classified as either present or not present because it is more a matter of extent rather than presence. In particular, Polya (1962) identified four levels of understanding a mathematical rule: (a) "mechanical"—a memorized method that can be applied correctly, (b) "inductive"—acceptance that explorations with simple cases extend to complex cases, (c) "rational"—acceptance of the proof of the rule as demonstrated by someone else, and (d) "intuitive"—personal conviction as to the truth beyond any doubt. Such levels characterize understanding as knowledge associated with mathematical rules. Similar characterizations occur in the writings of other researchers during this period. For instance, Flavell (1977) spoke of number knowledge or understanding in his discussions of number conservation whereas Davis (1978) argued understanding's dependence on whether the knowledge involved concepts, generalizations, procedures or number facts. Lehman (1977) equated understanding with three types of knowledge—applications, meanings, and logical relationships.

	Concrete	Iconic	Symbolic
Relational			
Instrumental			
Logical			

TABLE 1. Tall's categories of understanding

1.2. "Understanding" after 1978. Skemp's (1976) work distinguished understanding from knowledge and emphasized categories of mathematical understanding (Byers & Erlwanger, 1985). In particular, Skemp (1976) classified *relational* understanding as knowing what to do and why it should be done and *instrumental* understanding as having rules without the reasons. Each of these understandings has its own set of advantages. Instrumental understanding, according to Skemp (1976), tends to enable easy recall, to promote more tangible and immediate rewards, and to provide quick access to answers. Relational understanding, on the other hand, provides avenues for more efficient transfer, for extracting information from a learner's memory, for having such understanding as a goal in to itself, and for propagating the growth of understanding. Later, the classification evolved to include a third category denoted as *logical*—organization as in formal proof (1979) and finally a fourth category identified as *symbolic*—connection of symbolism and notation to associated ideas (Skemp, 1982), thereby creating four categories of understanding—relational, instrumental, logical, and symbolic—each subdivided into reflective and intuitive subcategories.

The categories of relational and instrumental understanding spawned a variety of other categorizations, namely (a) procedural and conceptual, (b) concrete and symbolic, and (c) intuitive and formal (Ball, 1991; Herscovics & Bergeron, 1988; Hiebert & Lefevre, 1986; Hiebert & Wearne, 1986; Nesher, 1986; Ohlsson, Ernest, & Rees, 1992; Resnick & Omanson, 1987; Schroeder, 1987). Byers and Herscovics (1977), combining the ideas of Bruner and Skemp, developed a tetrahedral classification of understanding with the following categories: instrumental, relational, intuitive, and formal. Tall (1978) suggested a matrix of categories (see Table 1) as characterizing understanding.

Other more general views propose that understanding is the development of connections between ideas, facts, or procedures (Burton, 1984; Davis, 1984; Ginsburg et al., 1992; Greeno, 1977; Hiebert, 1986; Hiebert & Carpenter, 1992; Janvier, 1987; McLellan & Dewey, 1895; Michener, 1978; Nickerson, 1985; Ohlsson, 1988). Forming a network of these connections provides a structure for situating new information by recognizing similarities, differences, inclusive relationships, and transference relationships between models. Thus, the development of understanding is a process of connecting representations to a structured and cohesive network. The connection process requires the recognition of relationships between the piece of knowledge and the elements of the network as well as the structure as a whole.

1.3. Recent views of "Understanding". Even though researchers now separate understanding from knowledge, evidence exists that the mathematics education community has not reached unilateral agreement as to the meaning of "understanding" since various authors approach it from diverse viewpoints (Schroeder, 1987). In particular, recent constructivist conceptualizations of understanding have been proposed in addition to Pirie and Kieren's model of the growth of mathematical understanding and Dubinsky's APOS theory. These include frameworks such as *cognitive* or *epistemological* obstacles (Bachelard, 1938; Cornu, 1991; Sierpinska, 1990b), *concept definition* and *concept image* (Davis & Vinner, 1986; Tall, 1989, 1991; Tall & Vinner, 1981; Vinner, 1983, 1991), *multiple representations* (Kaput, 1985, 1987a, 1987b, 1989a, 1989b) and a dichotomy between *operational* and *structural* conceptions (Sfard, 1991, 1992, 1994). Many of these characterizations have common elements especially since most derive from an underlying constructivist perspective that a learner's understanding is built by forming mental objects and making connections among them.

1.3.1. *Understanding as overcoming cognitive obstacles.* The concept of *cognitive* obstacles helps identify difficulties students encounter as they engage in the learning enterprise and therefore becomes useful in constructing better strategies for teaching (Cornu, 1991). *Cognitive* obstacles, first defined by Bachelard (1938), are classified as *genetic* or *psychological* obstacles, *didactical* obstacles, or *epistemological* obstacles depending on if they arose because of personal development, the teaching practice, or the nature of the mathematical concepts (Cornu, 1991). In particular, epistemological obstacles contain two essential attributes: (a) they are unavoidable as one constructs understandings of some mathematical concepts, and (b) the historical development of the concept reflects their existence (Cornu, 1991).

Sierpinska (1990b) drew her conceptualization of understanding from Lindsay, Husserl, Dilthey, Dewey, and Ricoeur and regarded "understanding as an act, but an act involved in a process of interpretation being a developing dialectic between more and more elaborate guesses and validations of these guesses" (p. 26). From this perspective, understanding derives its basis from the learner's ideologies, predispositions, preconceptions, convictions, and unperceived schemes of thought. This foundation may contain factors that act as obstacles to the further construction of understanding. Surmounting such an obstacle requires the learner to experience a mental conflict that calls into question the learner's convictions. Additionally, Sierpinska (1987) commented "if the presence of an epistemological obstacle in a student is linked with a conviction of some kind then overcoming this obstacle does not consist in replacing this conviction by an opposite one. This would be falling into the dual obstacle" (p. 374). As a result, overcoming an obstacle means that the learner must divest held convictions and analyze those beliefs from an external reference point. In doing this, the learner can recognize the tacit assumptions responsible for the cognitive dissonance and then evaluate alternative hypotheses. This evaluation requires the learner to identify the objects associated with a concept, identify common and disparate properties of the objects, generalize the scope of an concept's application, and finally synthesize the relationships between properties, facts, and objects to organize them into a consistent whole.

Not every act of understanding corresponds to an act of overcoming an epistemological obstacle but in general these can be equated. For instance, Sierpinska (1990a) stated "overcoming epistemological obstacles and understanding are two complementary pictures of the unknown reality of the important qualitative changes in the human mind. This suggests a postulate for epistemological analyses of mathematical concepts: they should contain both the 'positive' and the 'negative' pictures, the epistemological obstacles and the conditions of understanding" (p. 28). Thus, in Sierpinska's eyes, the use of an epistemological analysis of a mathematical concept helps determine a student's held understanding by making the observer aware of the various ways of perceiving a concept and the potential pitfalls inherent to them. The development of understanding is describable in many cases as the conscious awareness of an obstacle and this awareness leads to new ways of knowing. These new ways of knowing can result in the unfortunate acquisition of new epistemological obstacles. In particular, the act of overcoming an epistemological obstacle can merely open the learner to a larger domain containing additional epistemological obstacles. From this perspective, epistemological obstacles act as the dual to understanding since epistemological obstacles focus backwards on the errors and understanding looks forward to the new ways of knowing (Sierpinska, 1990b). Gauging the depth of understanding is accomplished through the identification of the number and quality of acts of understanding achieved or the number of epistemological obstacles overcome. These viewpoints provide complementary pictures of the qualitative changes in the mind as a learner interacts with concepts.

1.3.2. *Understanding as generating concept images and concept definitions.* According to Vinner (1991), learner's acquire concepts when they construct a *concept image*—the collection of mental pictures, representations, and related properties ascribed to a concept. Tall and Vinner (1981) write:

> We shall use the term *concept image* to describe the total cognitive structure that is associated with the concept, which includes all the mental pictures and associated properties and processes ... As the concept image develops it need not be coherent at all times ... We call the portion of the concept image which is activated at a particular time the *evoked concept image*. At different times, seemingly conflicting images may be evoked. Only when conflicting aspects are evoked *simultaneously* need there be any actual sense of conflict or confusion. (p. 152)

Evidently, a concept image differs from a concept's formal definition, if one exists, since a concept image exemplifies the way a particular concept becomes viewed by an individual (Davis & Vinner, 1986). The concept image involves the various linkages of the concept to other associated knowledge structures, exemplars, prototypical examples, and processes. As a result, the concept image is the overall cognitive structure constructed by a learner; however, in different contexts distinct components of this concept image come to the foreground. These excited portions of the concept image comprise the *evoked concept image* which consists of a proper subset of the concept image. This distinction between the image and the evoked image permits one to explain how students can respond inconsistently, providing evidence of understanding in one circumstance and a lack of understanding in another. A learner's description of his or her understandings may supply other discrepancies. In particular, any concept image has a related *concept definition*—the form or words used by a student to specify the concept. This concept definition, however, can differ from the formal mathematical definition of a concept since the concept definition is an individualized characterization of the concept.

Construction of these concept images occurs as a learner encounters new information and faces the consolidation of this information into the already present cognitive structure. The process of incorporation relates to the Piagetian notion of transition and in particular, assimilation and accommodation (Tall, 1991). *Assimilation* involves the taking in of new data and forming linkages between this new information and the original structure. In contrast, *accommodation* reorganizes part or the whole of the individual's cognitive structure. Underlying assimilation and accommodation are two essential mechanisms of cognitive development: generalization and abstraction. In mathematics, *generalization* typically refers to the process of applying an argument to a broader context; however, the type of generalization employed by a learner is dependent on the learner's already present cognitive structure. *Abstraction*, on the other hand, occurs when the learner focuses on the properties of an object and then considers those properties in isolation from the object from which they were derived. In this case, the structure of properties becomes an entity unto itself having application in other related domains.

Harel and Tall (1991) identified three types of generalization: expansive generalization, reconstructive generalization, and disjunctive generalization. *Expansive generalization* refers to the learner expanding the range of applicability of an existing schema[2] without reconstructing it. In other words, the learner sees previously applied methods as special cases of a new generalized procedure. In this case, the scope of application broadens without restructuring the internal cognitive structure even though the previous elements of the cognitive structure become substructures under the expanded scheme of application. *Reconstructive generalization* on the other hand occurs when the learner "reconstructs an existing schema in order to widen its applicability" (Harel & Tall, 1991, p. 38). That is, the learner searches for a new structure that takes initially isolated but related procedures or concepts and organizes the simpler elements as special cases under a more general case. *Disjunctive generalization* is perhaps the most worrisome from the prospective of teaching. In this type of generalization, the learner expands the scope of applicability by constructing a new, disjoint schema unconnected to previously studied concepts or procedures that could be considered special cases of this new schema. When speaking of disjunctive generalization, Tall (1991) stated "It is a generalization in the sense that the student may now be able to operate on a broader range of examples, but it is likely to be of little lasting value to the student as it simply adds to the number of disconnected pieces of information in the student's mind without improving the student's grasp of the broader abstract implications" (p. 12). As a result, disjunctive generalization encumbers the learner with an additional burden that can result in failure.

Abstraction differs from generalization in terms of the cognitive focus of the learner. Rather than extending ideas from one context to another, abstraction occurs when the learner focuses on the underlying structure of the contexts and extrapolates the common qualities or features. This process culminates in the construction of a set of axioms. As a result, abstraction requires a massive mental reconstruction in order to build the properties of the abstract object (Tall, 1991). Once the properties are abstracted, their application to a new context requires reconstructive generalization since "the abstracted properties are reconstructions of the original properties, now applied in a broader domain" (Harel & Tall, 1991, p. 39). In particular, extending the range of possible applications provides the opportunity for expansive generalization to transpire since the learner can relate the *abstract theory* (i.e. the axioms and their consequences) to elements in another cognitive structure. There is a problematic element, similar to compartmentalized

[2]Here, a schema refers to a structure developed by the learner to organize the various concepts, procedures, etc. linked to an endeavor such as solving linear equations in one variable.

knowledge from disjunctive generalization, associated with abstraction. If a learner encounters a limited set of examples, they may contain properties which do not hold for the entire class of objects. As a result, the learner must return, perhaps repeatedly, to reconstruct the abstract object and eliminate the nonessential properties. Even though these mechanisms may be problematic at times, their application permits the learner to build more extensive and possibly better interconnected, except in the case of disjunctive generalization, concept images.

1.3.3. *Understanding as operating with multiple representations.* According to Kaput (1989a), cognitive power exists in multiple, linked representations. These provide redundancy while permitting the learner to suppress some aspects of complex ideas and emphasize others. Facility with these representations and their linkages permits the learner to understand complex ideas in new ways and effectively apply them (Kaput, 1989a). The term *representation* is a trans-theoretic term for it encompasses a variety of characterizations: *cognitive* and *perceptual*, *computer*, *explanatory*, *mathematical*, and *symbolic* (Kaput, 1985, 1987b). In general, a *representation system* (or *symbol system*) aids in the instantiation of mathematical objects, relations, and processes by creating an environment in which shared cultural or linguistic artifacts can be expressed amongst the community (Kaput, 1989a). Such a representation system involves: (a) two worlds, the represented and the representing; (b) the elements of the represented world being represented; (c) the elements of the representing world doing the representing; and (d) the correspondence that affixes the connection between the two worlds (Kaput, 1985).

Mathematical representations or *mathematical symbol systems* are a special representation system where the represented world is a mathematical structure and the representing world is a *symbol scheme*[3] containing special correspondences (Kaput, 1987b). In particular, mathematical symbol systems, similarly to natural language and pictorial systems, manage the influx of experience by breaking it into chunks, assigning symbols to those chunks, and coordinating these notations in a milieu devoid of the original nuances and complex referents (Kaput, 1987b). The mathematical symbol system and its connections then can form a structure which acts as a symbol system used to represent another symbol system thereby exhibiting a self-similarity under magnification.

Kaput (1989a) contends that a learner's development and expression of mathematical meaning can be viewed from the lens of the construction of notational forms and structures in mental representations where a *mental representation* is the means by which an individual organizes and coordinates the flow of experience. In particular, Kaput (1987a) stated "The fundamental premise is that the root phenomena of mathematics learning and application are concerned with representation and symbolization because these are at the heart of the content of mathematics and are simultaneously at the heart of the cognitions associated with mathematical activity" (p. 22). In this view, the growth of mathematical meaning is the construction and utilization of representations and symbolization.

[3]"A symbol scheme is a concretely realizable collection of characters together with more or less explicit rules for identifying and combining them" (Kaput, 1987b). In essence, a symbol scheme is an abstract means of representing a more complex concepts using quasi-realistic symbols to delineate typically intangible objects. For instance, the Hindu-Arabic numbers with their concatenations would satisfy the definition of a symbol scheme. Similarly, the coordinate axes with their syntactical rules would also be considered a symbol scheme.

As a learner constructs personal meaning, negotiation occurs between two separate worlds: *physical operations* which are observable and *mental operations* which are hypothetical. In particular, the development of understanding is the movement from operating in the world of physical operations to operating in the world of mental operations. In order to accomplish this, the learner must employ "(i) deliberate, active interpretation (or 'reading'), and (ii) the less active, less consciously controlled and less serially organized processes of having mental phenomena evoked by physical material" (Kaput, 1992, p. 522). Underlying the physical operations are the notations used to display the operations. A *notation system*[4] is separate from any particular physical representation (thereby differentiating it from a symbol scheme which must be linked to two worlds: the represented and the representing) and contains a set of rules that define its objects and the allowable actions upon those objects.

With this in mind, one can recognize that most true mathematical activity involves the *coordination of* and *translation between* different notation systems (Kaput, 1992). Given two notation systems with their sets of rules defining the objects of the system, the allowable actions upon those objects, and in the case of symbol schemes the correspondent physical mediums in which the systems may be instantiated, a learner can (i) negotiate between the two notation systems, (ii) integrate cognitions, (iii) transform objects within a particular representation, or (iv) cognize about a notational system. The first two mechanisms comprise *referential extensions* since they "horizontally translate" between either notational systems or mathematical structures. In the first case, meaning arises from identifying the connected components of different representational systems through translation and in the second, through the construction and testing of mathematical models, which amount to translation and coordination of cognitive organizations.

The latter two mechanisms amount to *consolidations* whereby the learner engages in "vertical growth" by transforming actions at one level into objects and relations capable of being operated upon at another level. The third mechanism produces meaning through pattern and syntax learning through transformations within a particular notation system that may or may not contain references to external meanings. This growth is a reorganization of the notational system creating a hierarchical structure albeit within the original notation system. In contrast, the last mechanism, cognition about a notational system, yields mathematical meaning by reifying[5] actions, procedures, and concepts into phenomenological objects that can potentially act as the basis for new actions, procedures, and concepts as a higher level of organization (Kaput, 1992). These mechanisms produce meaning and therefore develop understanding for the learner by either creating new linkages between representation systems or reorganizing elements within a representation system.

1.3.4. *Understanding as constructing operational and structural conceptions.* Sfard (1991) defines the building blocks of mathematics as two entities: *concept* and *conception.* A concept refers to an official mathematically defined idea whereas

---

[4]Familiar notation systems, such as numeration systems and algebraic notation systems for one or several variables, typically include textual elements but they can also include strictly pictorial elements correspondent to Dienes blocks, Cuisenaire rods, fraction bars, etc.

[5]Reification, a Piagetian construct used also by Sfard and APOS theory, is the ability of the learner to envision, almost simultaneously, the results of processes as permanent objects inseparable from the underlying processes from which they arose.

a conception involves a cluster of the learner's internal representations and linkages caused by the concept. These two definitions are related to Vinner's (1991) discussion of formal concept definition and the description by Tall and Vinner (1981) of a concept image. Sfard (1991) asserts that mathematical concepts inhabit a duality of conception for they can be viewed as static, instantaneous and integrative—*structural* or dynamic, sequential, and detailed—*operational*.

An operational conception, although difficult to describe, concerns processes, algorithms, and actions which occur at the physical or mental level. A *structural* conception on the other hand is more abstract, more integrated, and less detailed than an operational conception. In particular, a structural conception is somehow isomorphic to the ability to "see" advanced mathematical constructs which are not physical entities but rather abstract mental organizations perceivable in only one's mind's eye (Sfard, 1991). This capability of seeing the invisible objects that form the mathematical concept draws one into the world of visualization. Sfard proposes that structural conceptions receive support from compact and integrative mental images rather than from verbal representations which require serial processing. These mental images permit the learner to make the abstract ideas more tangible and conceive them as almost physical entities where the operations upon them occur entirely in the mind's eye. In addition, such visualization empowers the learner to develop a holistic view of the concept thereby allowing observations from multiple perspectives while preserving the identity of and relationships within the concept.

From Sfard's perspective, understanding reaches beyond an ability to solve problems or prove theorems (Sfard, 1994). In essence, she equates understanding with the construction of links between symbols and the development of a structural conception. However, according to Sfard (1991), "*there is a deep ontological gap between operational and structural conceptions*" (p. 4). Even though a gap exists, operational and structural conceptions are not mutually exclusive. In particular, they are complementary in the sense that they are two views of the same mathematical concept and are inseparable since the concept harbors both operational and structural elements. The operational conception views the concept as a process and the structural conception equates the concept with a static object transcendent of its process roots. However, for conceptual development, both of these conceptions are necessary. For instance, Sfard (1991) states "Indeed, in order to speak about mathematical objects, we must be able to deal with *products* of some processes without bothering with the processes themselves ... It seems, therefore, that the structural approach should be regarded as the more advanced stage of conceptual development. In other words, we have good reasons to expect that *in the process of concept formation, operational conceptions would precede the structural*" (p. 10). This statement does not imply that structural conceptions can only develop after the construction of operational conceptions, rather that in general this is the natural path for development. In other words, as one views the historical development of many mathematical concepts, society moved through a series of stages culminating in a structural conception. In particular, Sfard (1991) used a historical analysis of concept formation to identify three distinct stages in the process: the generation of a process from already familiar objects, the emergent recognition of the processes as autonomous entities, and the ability to conceive the new entity as a synthesized, object-like structure.

These three stages are classified as *interiorization*, *condensation*, and *reification*, respectively. Interiorization occurs as a learner becomes familiar with the processes which eventually can be reified into a mathematical object. Interiorization in this context remains similar to the mechanism described by Piaget (1970) for it essentially comprises a movement from conceptions based upon physical operations to those founded on mental representations of the processes. The everyday meaning of condensation is similar to its meaning in this context. Rather than working through a long sequence of related but distinct mental processes, condensation enables the learner to conceive of a sequence as a single process relating input and output without the intervening steps. The attachment of a name to the condensed sequence gives birth to a new concept which remains affixed to a process orientation until it becomes reified.

Reification, according to Sfard and Linchevski (1994), is ultimately responsible for the development of mathematical objects. In essence, reification is a quantum leap from conceiving the new entity as tightly connected to a process to conceiving the notion of the entity as an object which itself can be acted upon. As a result, reification is an ontological shift on the part of the learner (Sfard, 1991). This shift permits the capability to see something familiar from an entirely different perspective that detaches the condensed sequence from the originating sequence. The ensuing structure, although invariably connected to the processes it exemplifies, can now be viewed as a static object in the mind's eye. In addition, the new entity begins to draw its meaning from its membership not in the realm of processes but rather as a member of a category of abstract objects which enhances the scope of applications. In particular, Sfard (1991) declares:

> At some point, this category rather than any kind of concrete construction becomes the ultimate base for claims on the new object's existence. A person can investigate general properties of such category and various relations between its representatives. He or she can solve problems involving finding all the instances of the category which fulfill a given condition. Processes can be performed in which the new-born object is an input. New mathematical objects may now be constructed out of the present one. (p. 20)

Once a process has been reified, it yields an object upon which a higher level process can act. This process can then become interiorized and the entire cycle repeated. As a result, the three-phase system of interiorization, condensation, and reification is generally hierarchical and repetitive.

The development of understanding in this view is the capability to break free from one's own constrained thinking. In doing so, the learner gains the capability to perceive a process no longer as a sequence of physical acts which have been interiorized and condensed but rather as an object. This object along with the other objects in the mathematical universe and the operations potentially performed upon them, according to Sfard (1994), receive meaning though metaphorical reflection. In particular, Sfard (1994) aligns herself with Lakoff and Johnson's thesis that "metaphors constitute the universe of abstract ideas, that they create rather than reflect it, that they are the source of our understanding, imagination, and reasoning" (p. 47). The metaphor therefore provides meaning to an abstract concept since it provides a figurative projection of operations performed in a physical reality into the world of ideas. This characterization remains true even when the concept appears far removed from physical reality since underlying the concept is a long chain of metaphors eventually rooted in actions performed in a physical reality. As a result, it is Sfard's contention that reification, the transition from an operational to a structural mode of thinking, accounts for the construction of mathematical concepts and this fabrication corresponds to the birth of a metaphor. This metaphor brings the mathematical object into being and thereby deepens the learner's understanding of the mathematical universe.

2. Pirie and Kieren's Model of Understanding

Pirie and Kieren consider their work on the growth of mathematical understanding to provide some answers to the cogent questions raised by Sierpinska (1990b): "Q1. Is understanding an act, an emotional experience, an intellectual process, or a way of knowing? ... Q3. Are there levels, degrees or rather kinds of understanding? ... Q5. What are the conditions for understanding as an act to occur? ... Q7. How do we come to understand? ... Q8. Can understanding be measured and how?" (p. 24). Pirie and Kieren's initial definition of mathematical understanding evolved from Glasersfeld's constructivist definition of understanding (Kieren, 1990). In particular, Glasersfeld (1987) proposed the following definition of understanding:

> The experiencing organism now turns into a builder of cognitive structures, intended to solve such problems as the organism perceives or conceives ... among which is the never ending problem of consistent organizations [of such structures] that we call understanding. (p. 7)

Glasersfeld perceived understanding as the continual process of organizing one's knowledge structures. Using this definition as an advance organizer, Pirie and Kieren (1991a) began to develop their theoretical position concerning mathematical understanding. They characterize mathematical understanding as the following:

> Mathematical understanding can be characterized as leveled but non-linear. It is a recursive phenomenon and recursion is seen to occur when thinking moves between levels of sophistication. Indeed each level of understanding is contained within succeeding levels. Any particular level is dependent on the forms and processes within and, further, is constrained by those without. (Pirie & Kieren, 1989, p. 8)

It is the purpose of the following discussion to explicate Pirie and Kieren's model more precisely.

2.1. Elements of Pirie and Kieren's Model of the Growth of Mathematical Understanding. Using the above definition, Pirie and Kieren conceptualize their model of the growth of mathematical understanding as containing the eight potential layers shown in Figure 1. The process of coming to understand begins the core of the model called the *primitive knowing*[6] layer. Primitive connotes a starting place rather than low level mathematics. The core's content is all the information brought to the learning situation by the student. These contents have been discussed under various names: "intuitive knowledge" (Leinhardt, 1988),

---

[6]The primitive knowing layer was initially referred to as "primitive doing" or "doing" in articles prior to 1991.

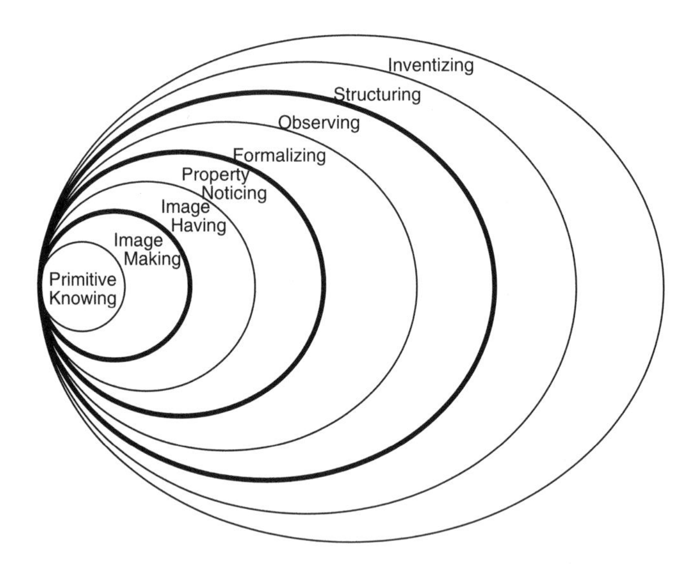

FIGURE 1. A diagrammatic representation of the model for growth of mathematical understanding

"situated" knowledge (Brown, Collins, & Duguid, 1989), and "prior" or "informal" knowledge (Saxe, 1988). For a particular concept such as fractions, one can surmise that in tracing the growth of mathematical understanding a learner comes to the learning situation with a host of information which may or may not inform the growth of understanding. Resnick and Omanson (1987) noted that students entered a mapping instruction[7] with mental representations of block subtraction attached to a rich knowledge base associated with subtraction. Here the learner brings both understandings of subtraction beyond the primitive knowing layer and other understandings supportive of continued growth. In contrast, Hiebert and Wearne (1986), in examining decimal learning, found students to perceive decimal symbols as part of a new symbol system accompanied by a new set of rules thus minimizing the links to previously learned material. In turn, the core's under-utilized links can impact the process of understanding.

At the second layer, called *image making*, the learner is able to make distinctions based on previous abilities and knowledge. These images are not necessarily "pictorial representations" but rather convey the meaning of any kind of mental image. The actions at this layer involve the learner doing, either mentally or physically, something to gain an idea about a concept. For instance, the learner while engaged in folding or cutting activities may develop an image of fractions as things gained from cutting something into equal smaller pieces. As a result, the actions at this layer involve the development of connections between referents and symbols as described by Wearne and Hiebert (1988), Greeno (1991) and Brownell (1945) as the learner employs fraction language to discuss and record the actions. At the next layer, called *image having*, single-activity associated images are replaced by

[7]In this case, the mapping instruction involved three components: (1) learning and practicing subtraction with both base 10 blocks and symbols, (2) practicing symbol recording without physically moving the blocks, and (3) doing the written symbol manipulation without the presence of the blocks.

a mental picture. The development of these mental pictures, or more precisely, mental process-oriented images, frees the learner's mathematics from the need to perform particular physical actions (Pirie & Kieren, 1992b). These mental images have been discussed under the guise of several names: "concept image" (Davis & Vinner, 1986), "frames" and "knowledge representation structures" (Davis, 1984), and "students' alternative frameworks" (Driver & Easley, 1978). The freedom to imagine a concept unconstrained by the physical processes which elicited the image is useful to the growth of mathematical knowing since the learner begins to recognize obvious global properties of the inspected mathematical images.

At the fourth layer referred to as *property noticing*, the learner can examine a mental image and determine the various attributes associated with the image. Besides noticing the properties internal to a specific image, the learner is capable of noticing the distinctions, combinations or connections between multiple mental images. These properties combine to construct evolving definitions that may identify particular characteristics while ignoring other elements of the concept. According to Pirie and Kieren (1991a), the difference between image having and property noticing is the ability to notice a connection between images and explain how to verify the connection. These connections, according to Michener (1978), arise from the exploration and manipulation of a concept at many levels such as "examining relevant examples, perturbing settings and statements and fiddling around numerically and pictorially" (p. 373). It may be at this property noticing layer that the learner notices commonalties of various images and develops a concept definition (Tall & Vinner, 1981) built upon the interplay between multiple linked images rather than disconnected images. In reference to understanding fractions, the actions of the learner reveal a recognition that equivalent fractions are generated by multiplying numerator and denominator by the same factor. The verbalization associated with this layer would consist of producing a series of equivalent fractions such as

$$2/3 = 4/6 = 6/9 = \dots \text{ (Pirie \& Kieren, 1991a, p. 2)}$$

The difference between the property noticing layer's actions and those of image having layer is a capability to notice equivalencies and explain the necessary techniques for developing them.

At the fifth layer of understanding, called *formalizing*, the learner is able to cognize about the properties to abstract common qualities from classes of images. At this layer the learner has class-like mental objects built from the noticed properties, the abstraction of common qualities, and abandonment of the origins of one's mental action (Pirie & Kieren, 1989). Description of these class-like mental objects results in the production of full mathematical definitions. The language used to describe a concept does not have to be formal mathematical language; however, the general descriptions provided by students must be equivalent to the appropriate mathematical definition. It may be at this layer that Michener's description (Michener, 1978) of the first of a three-phase passage to full understanding comes into play. This first stage consists of the learner gaining familiarity with the concept and its neighboring concepts. Acquisition of the definitions occurs but the concern is with the particular concept and the scope of connections between it and other concepts remains minimal and local. With respect to fractions, the learner can now speak of fractions as a class of formal objects unconnected to specific examples and represent this class in terms of a/b at the formalizing layer. The learner also can view fractions as a set of numbers or general entities no longer action-oriented.

The following layer of understanding, *observing*, entails the ability to consider and reference one's own formal thinking. Beyond the learner engaging in metacognition, the learner is also able to observe, structure and organize personal thought processes as well as recognize the ramifications of thought processes. At this layer, the learner can produce verbalizations concerning cognitions about the formalized concept. This description echoes the second stage proposed by Michener (1978) in which the learner gains an overall conceptualization of the subject matter and its development. In particular, the concern resides on "items and relations within the representation spaces and the theory as a whole; it is more global in outlook" (Michener, 1978, p. 376). The learner combines the definitions, examples, theorems, and demonstrations to identify the essential components, connecting ideas, and the means for traversing between those ideas. For the case of fractions, the learner has progressed to the production of verbalizations concerning cognitions connected with fractions at the organizing layer. Here, the learner might observe that: "There can be no smallest half fraction" (Pirie & Kieren, 1992b, p. 247). Such an observation is different from the property noticing layer recognition, any half fraction can be made smaller because it can be folded in half, or the image having layer realization of many folds produces small pieces.

Once one is capable of organizing one's formal observations, the natural expectation is to determine if the formal observations are true. The learner after gaining such awareness can then explain the interrelationships of these observations by an axiomatic system (Pirie & Kieren, 1989). This layer is called *structuring*. At this layer, the learner's understandings transcend a particular topic for the understanding inhabits a larger structure. This structuring layer appears well correlated with the third stage described by Michener (1978). In this third stage, the learner begins to see relationships between several subjects, address certain questions about the underlying ideas, axioms and examples, relate these underlying ideas across multiple domains, and perceive the interconnectedness of several theories. Therefore at the structuring layer, fractions are conceived as beyond the physical entities associated with the image making layer, the action-oriented equivalencies associated with the property noticing layer, and the resultant of formal algorithms associated with the formalizing layer. The learner would now be able to conceive proofs of properties associated with fractions such as the closure of half fractions under addition where the addition of fractions is viewed as a logical property following from other logical properties (Pirie & Kieren, 1991a, 1992b).

The outermost ring of the Pirie and Kieren's model of mathematical understanding is called *inventizing*. Originally referred to as *inventing*, this layer's name changed to the present term to distinguish the activities associated with this layer and the inventions that can occur at lower layers of understanding (Pirie & Kieren, 1994b). As a result, the use of inventizing does not imply that one cannot invent at other layers, but rather is used to indicate the ability to break free of a structured knowledge which represents complete understanding and to create totally new questions which will result in the development of a new concept. At this layer, the learner's mathematical understanding is unbounded, imaginative and reaches beyond the current structure to contemplate the question of "what if?" This questioning results in a learner's use of structured knowledge as primitive knowing when investigating beyond the initial domain of exploration. For example, extension of the fraction notation a/b for $a + bi$ to $a/b/c/d$ for $a + bi + cj + dk$ took the mathematician Hamilton from a structured understanding of complex numbers into a new system called the quaternions (Pirie & Kieren, 1991a).

2.2. Constructs of Pirie and Kieren's Model of Understanding. Pirie and Kieren's model of understanding contains an inherent dynamism which is apparent in several components. In particular, the core of the model, primitive knowing, has an underlying dynamic quality. Pirie and Kieren (1990), for instance, characterize this movement in the following statement:

> One obvious consequence of this model is that, outer levels grow recursively from the inner ones, but knowing at an outer level allows for and indeed retains the inner levels. Outer levels embed and enfold the inner ones. In fact this [is] a relativity theory of understanding and therefore a particular feature of primitive doing [referred to in later papers as primitive knowing] is that observers can consider what they wish as the focus of this level. For example, one could observe a person at the inventing [later referred to as inventizing] level as having their entire previous understanding as a new primitive doing [primitive knowing]. A main consequence of this line of thinking is that, to an observer, understanding has a fractal quality. One could look at the understanding of a person "inside" primitive doing [primitive knowing] and observe the same leveled structure. (p. 5)

From this, one can see that Pirie and Kieren view the inner core, called primitive knowing, as composed of complete models similar to the whole. This property gives the inner core the attributed *fractal characteristic* evidenced in Figure 2. This nesting points to the importance of the information contained in the inner core since that information constrains one's knowing at outer layers (Pirie & Kieren, 1989). As a result, this core knowledge can either beneficially aid a student in understanding a concept or hinder the student from understanding by acting as an obstacle (Mack, 1990; Resnick et al., 1989).

The most critical feature of Pirie and Kieren's model of understanding is the dynamic process of *folding back*. When one encounters a problem whose solution is not immediately attainable, one must see the necessity to fold back to an inner layer to extend one's current, inadequate understanding. The process of folding back to an inner layer finds one examining the layer's understandings in a manner different from the actions originally displayed when operating at the layer. This difference is both qualitatively and intently different due to the motivation associated with the folding back and the developed understandings of the outer rings (Pirie & Kieren, 1991a, 1992a). As a result, the extension of one's understanding is not simply a product of generalizing a given layer's activities nor a consequence of reflectively abstracting one's understanding to attain a new outer layer, but more usually the extension occurs by folding back to recursively reconstruct and reorganize one's inner layer knowledge, and so further extend outer layer understanding. Such a process is similar, according to Pirie and Kieren (1992b), to the reconstruction of an existing schema described both by Sfard (1991) and Harel and Tall (1991).

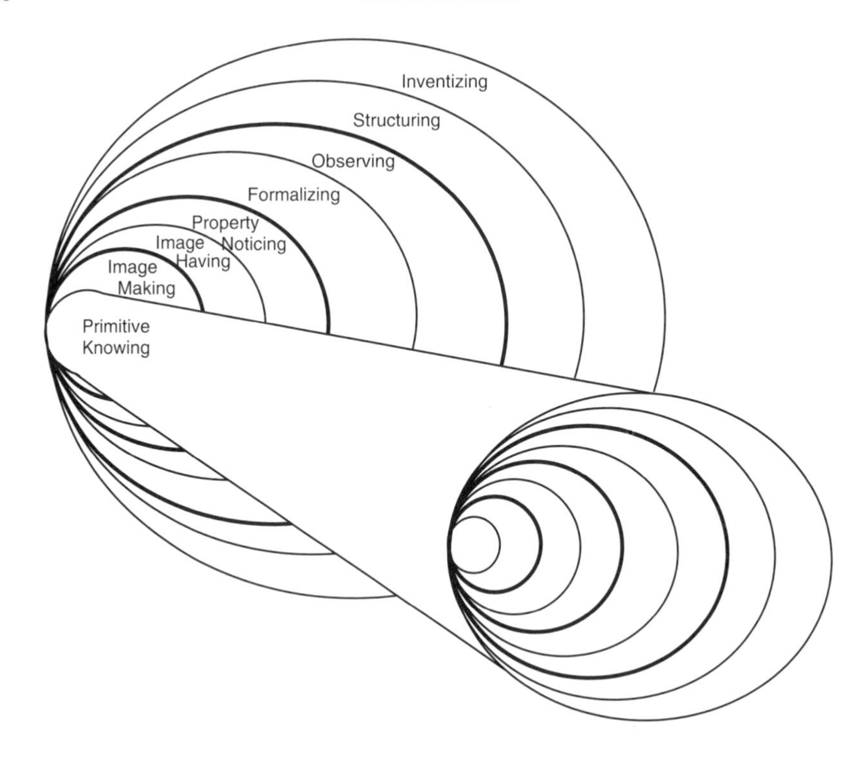

FIGURE 2. A diagrammatic representation of the model for growth of mathematical understanding illustrating the self-similar nature of the inner core called primitive knowing

Another construct of the model identifies the *complementarity of a process and a form-oriented action*. Each of the layers, beyond the primitive knowing layer, contains a complementarity of form and process as identified in Figure 3. Pirie and Kieren contend that one must exhibit the form-oriented action to be fully functioning at a layer (Kieren, 1990; Pirie & Kieren, 1990, 1994b). This form-oriented action acts as a demonstration to an external agent attempting to determine the layer of understanding at which a learner is operating. Therefore, the absence of a layer's displayed complementary action does not demonstrate that the learner is operating at a particular layer. Pirie and Kieren (1991a) extended these notions and re-labeled the diagrams shown in Figure 4 to allow easier discussion of the blended, laminar layers and the acting and expressing complements therein.

In particular, Pirie and Kieren (1994b) assert that if students perform only actions without the correspondent expression then their understandings are inhibited from movement to the next layer.

The image making layer is composed of two complementary elements called *image doing* and *image reviewing*. The image doing learner sees previous work as complete and does not return to it; whereas, an image reviewing learner involves the constructive alteration of previous behavior without necessarily noticing a pattern (Pirie & Kieren, 1991a). Image doing initially may appear ill-defined since engagement in any activity appears to be image doing. However, image doing, according

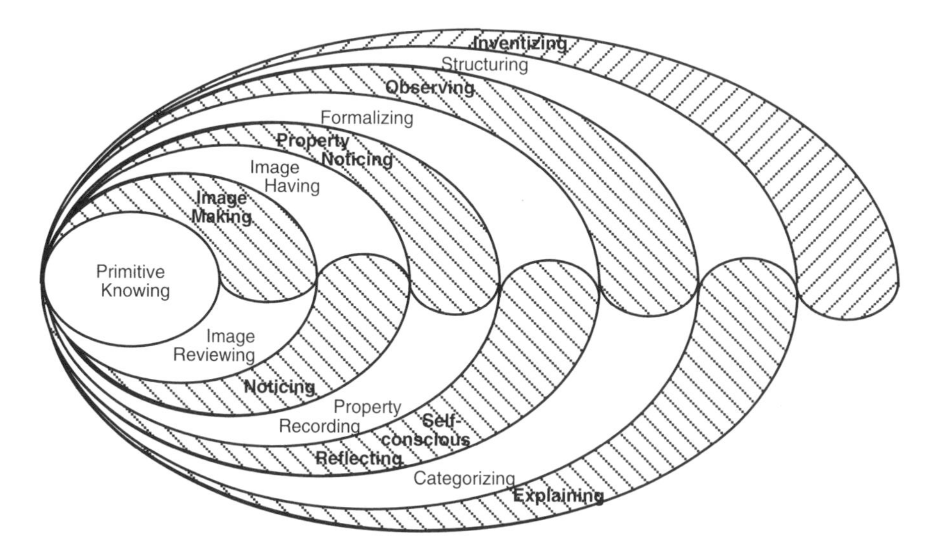

FIGURE 3. The within level complementarities

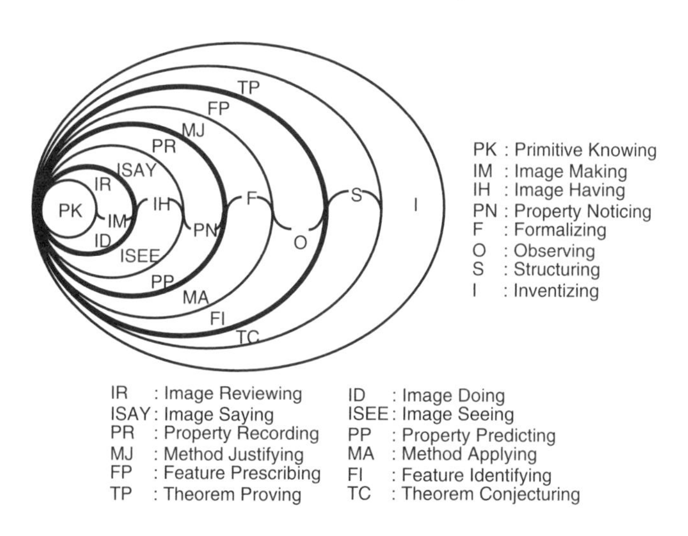

FIGURE 4. Rings with acting and expressing complements identified

to Pirie and Kieren (1994b), consists of only potentially fruitful actions linked to intentional making of some sort of images for a concept.

The image having layer has two complementary elements, *image seeing* and *image saying*. The *image seeing* act has collected together previously encountered examples and has a pattern; whereas, the *image saying* behavior articulates the pattern associated with the image (Pirie & Kieren, 1991a). In particular, when engaged in image seeing, a learner identifies a discrepant element as uncorrelated with the learner's mental image without being able to articulate why. On the other hand, image saying involves the learner in both articulating both the image and the reason the discrepant element does not fit the image (Pirie & Kieren, 1994b). At the property noticing layer, the two complementarities are *property predicting* and *property recording*. The *property predicting* act relates the image to a property noticed by the learner and *property recording* is an act of incorporating into the learner's cognitive structure the noticed property as something that exists and seems to work. According to Pirie and Kieren (1994b), both the image having and the property noticing layers have a particular characteristic that distinguishes them from other layers. At these two layers, the "acting" notions produce temporal understandings in the sense that they will diminish over time if not coordinated with their complementarity "expressing" notions. As a result, the lack of an "expressing" activity seems to obstruct movement beyond their previous images and therefore to higher layers in their model (Pirie & Kieren, 1994b).

At the formalizing layer, *method applying* and *method justifying* are the two complementarities whereas the observing layer contains the complementarities of *feature identifying* and *feature describing* (Pirie & Kieren, 1994b). The last layer to contain complementarities is the structuring layer and it contains *theorem conjecturing* and *theorem proving* (Pirie & Kieren, 1994b). The complementarities for these last three layers of their model are defined without any further description other than the provision of illustrative terms in Pirie and Kieren (1994b).

The last construct of the model is the darker rings called "*don't need*" boundaries. These boundaries signify movement of the learner to a more elaborate and stable understanding which does not necessarily require the elements of the lower layers (Pirie & Kieren, 1992b). For example, once students have moved to the image having layer, it is no longer necessary to elicit examples of image making or elements from the primitive knowing layer. A person at the image having layer is at a qualitatively different layer of understanding when actively involved in *image seeing* and *image saying* activities for the learner does not necessarily see a mathematical object as the result of a doing activity, but rather as an entity with identifiable features (Pirie & Kieren, 1991a). Movement from the image making layer to the image having layer involves a qualitative change in the associated thinking processes. The learner has moved from layers associated with unselfconscious knowing to conscious thinking. Therefore, moving over a "don't need" boundary signifies a major qualitative change in one's understanding. Similar qualitative changes have occurred when one achieves either the formalizing or structuring layers since there is no longer a need for an image or a concrete meaning for the mathematical activity. However, the overcoming of a "don't need" boundary does not imply that a learner may never pass back into that lower level understanding. In fact, these "don't need" boundaries are typically crossed back over during times of folding back in order to reorganize and reconstruct lower level understandings in order to expand outer level understandings.

3. Dubinsky's APOS Theory

Piaget's proposal of the process of *reflective abstraction* as the key to the construction of logico-mathematical concepts influenced Dubinsky's development of the Action-Process-Object-Schema theory (APOS theory). In this theory, the development of understanding " ... begins with manipulating previously constructed mental or physical objects to form actions; actions are then interiorized to form processes which are then encapsulated to form objects. Objects can be de-encapsulated back to the processes from which they were formed. Finally, actions, processes and objects can be organized in schemas" (Asiala et al., 1996, p. 8). The mechanism

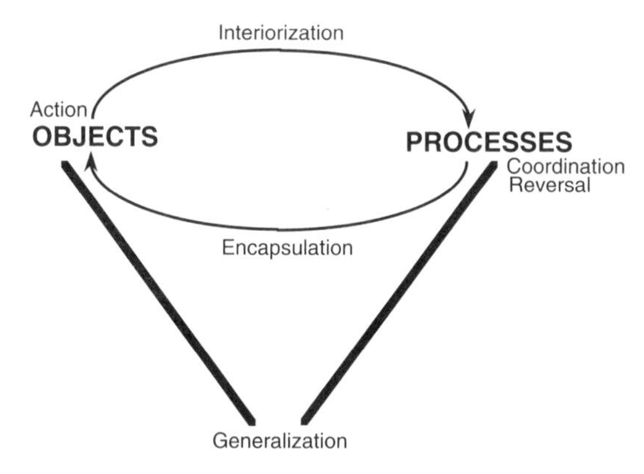

FIGURE 5. Schemas and their construction

of constructing these schemas, reflective abstraction, is the heart of APOS theory since it separates the properties connected to a concept and identifies the salient elements which comprise the concept apart from the context. In particular, reflective abstraction extends the construction of connections between abstracted concepts and builds a structure out of the related abstractions.

According to Dubinsky (1991), five different kinds of Piagetian constructions are essential to developing abstract mathematical concepts—*generalization*, *interiorization*, *encapsulation*, *coordination* and *reversal*. The fifth construction, *reversal*, considered to be crucial to advanced mathematical thinking from the APOS perspective was not part of Piaget's description of reflective abstraction although it was contained in his writings (Dubinsky, 1991). The ensuing sections describe APOS theory's components, define its various constructions, and relate the components and constructions.

3.1. Elements of APOS theory. Dubinsky proposes that a schema is more than a static entity since it remains inseparable from its own dynamic, continuous evolution. Figure 5 depicts the structure of a schema and identifies the influential constructions.

The primary building block of understanding is *action* (similar to Piaget's *action schemes*). Asiala et al. (1996) identify that understanding of a mathematical concept originates through the manipulation of previously constructed mental or physical objects to form actions. An action equates with any repeatable physical or mental operation that transforms either a physical or mental object in some manner. As a result, the actions tend to be algorithmic in nature and externally driven (Clark et al., 1997). In reference to the function concept, a learner may conceive a function as involving the plugging of numbers into an algebraic expression and calculating the resultant values. Such a conceptualization appears restricted and static since it remains inexorably linked to individualized evaluations that occur one at a time.

Even though an action conception appears limited in scope, it is a necessary element in the construction of understanding since actions are done on the mathematical objects within a learner's realm of experience. As an action becomes *interiorized* through a sequence of repeating the action and reflecting upon it, the action no longer remains driven by external influences since it becomes an internal construct called a *process* (similar to Piaget's *operations*). Attainment of this process conception indicates the learner can reflect on the process, describe it, and even reverse the steps of transformation without resorting to external stimuli (Asiala et al., 1996). In particular, the learner now can use this process to obtain new processes either through *coordination* or *reversal*. The coordination of multiple related processes results in the construction of new processes. Many times this is necessary as a learner encounters elements of a new topic and recognizes underlying structures that permit the application of several processes developed in a different context. Additionally, situations exist where the learner encounters topics which require the composition of seemingly unrelated processes to build a more complex structure (Dubinsky, 1991). Reversal, on the other hand, permits the learner to conceive of a new process that undoes the sequence of transformations comprising the initial process. Essentially, reversal is the construction of a process which countermands an internalized process. For instance with respect to the function concept, a learner can link two or more processes together to produce a composite function thereby coordinating several processes into a singular process or reverse a function to obtain the inverse of the function. In addition, the strengthening of a process conception permits greater access for the learner to notions such as one-to-one and onto.

When a learner can reflect on a process and transform it by some action, then the process is considered to have been *encapsulated* to become an *object*. According to Sfard and Linchevski (1994), the process of reification is virtually synonymous with encapsulation. Once encapsulated, the object exists in an individual's mind necessitating the assignment of a label to the object (Dubinsky et al., 1994). The resultant label allows the learner to both name the object and connect that name with the process from which the object was constructed. This dual vision of the object is essential because the learner needs to be able to *de-encapsulate* an object thereby returning to the process in a form prior to its encapsulation. De-encapsulation enables the learner to use the properties inherent in the object to perform new manipulations upon it. For instance, with respect to functions, a learner must be facile in encapsulating processes into objects and de-encapsulating objects into processes when considering manipulations such as adding, multiplying, or creating sets of functions (Asiala et al., 1996).

The final element of the APOS theory, the *schema*, is a collection of processes and objects (Dubinsky et al., 1994). This collection may be more or less coherent but the learner utilizes it to organize, understand, and make sense of observed phenomena or concepts. As a result, a schema generally contains other subordinate schemas to span across a particular domain. The term schema has strong connections to Piaget's use of the term *schemata* which in turn relates to Tall and Vinner's descriptions of a concept image (Asiala et al., 1996). However, the schemas designated by Dubinsky et al. (1994) correspond to Piaget's thematized schematas which indicates that the collection coalesced into an object upon which actions can take place. In particular, these schemas are the mental representations of concepts as they exist in the learner's mind. As a result, Dubinsky et al. (1994) point out that "A schema can be used to deal with a problem situation by unpacking it and working with the individual processes and objects. A schema may also be treated as an object in that actions and processes may be applied to it" (p. 271). This ability to conceive a schema as an object upon which actions and processes can transform the schema infers a fractal quality with schemas containing other schemas as elements.

In particular, schemas are organizing structures that incorporate the actions, processes, objects and other schemas that a learner invokes to deal with a new mathematical problem situation. Constructing such structures requires a mechanism called *generalization* which permits a wider scope of schema utilization also referred to as *expansive generalization* by Harel and Tall (1991). In particular, the generalization of an already constructed schema occurs when the learner enlarges the venue of applicability without changing the structure of the schema. Piaget referred to such a construction as a reproductive or generalized assimilation and called the generalization *extensional* (Dubinsky, 1991). Such generalization is the simplest and most familiar form of reflective abstraction since it involves the application of an already existent schema to a new set of objects.

More recently, Clark et al. (1997), while investigating student construction of the chain rule understanding, attempted to utilize this notion of schema to negotiate between members of the RUMEC research team explanations of the exhibited student understandings. However, it became apparent during their attempt to classify student responses that the descriptions of actions, processes, and objects were insufficient in this context. In response, Clark et al. (1997) state "In the course of these negotiations, we realized that Actions, Processes, and Objects alone were insufficient to describe the student understanding of the chain rule" (p. 353). In essence, for the concept of chain rule that incorporates a variety of sub-concepts, APOS theory did not provide a sufficiently robust explanation for all the various nuances encountered, since it does not result from the encapsulation of a single process.

Clark et al. (1997) then returned to the writings of Piaget and Garcia to examine a triad mechanism useful in the description of schema development. This mechanism defines three particular stages in the development of a concept: *Intra*, *Inter*, and *Trans*. "The Intra stage is characterized by the focus on a single object in isolation from other actions, processes, or objects. The Inter stage is characterized by recognizing relationships between different actions, processes, objects and/or schemas. We find it useful to call a collection at the Inter stage of development a pre-schema. Finally the Trans stage is characterized by the construction of coherent structure underlying some of the relationships discovered in the Inter stage of development" (Clark et al., 1997, pp. 353–354). This refinement of the concept of schema permitted Clark et al. (1997) to explain the actions and statements of students with respect to the chain rule as well as McDonald, Mathews and Strobel (2000) to study the cognitive development of the concept of sequence. Consequently, the notion of schema continues to be reexamined and redeveloped as the theory evolves.

3.2. Constructs of APOS Theory. The construction of understanding, from the APOS perspective, passes through several stages driven by external cues which then become internalized, reflected upon and eventually organized. In particular, Asiala et al. (1996) characterized the development of mathematical knowledge from the perspective of APOS theory in the following manner:

> An individual's mathematical knowledge is her or his tendency to respond to perceived mathematical problems situations by reflecting on problems and their solutions in a social context and by constructing or reconstructing mathematical actions, processes

> and objects and organizing these in schemas to use in dealing with the situations. (p. 7)

The act of reflection, a construct essential to APOS theory, is the learner's conscious attention to the operations being performed. This reflection plays an integral role in learning and knowing since it involves reaching beyond contemplation on the particular performance of techniques and algorithms no matter how complicated they may be (Asiala et al., 1996). In particular, reflection provides the learner with an awareness of how procedures work, a feeling for the results without physically performing the operations, an ability to analyze and manipulate variant algorithms, and a capability to see relationships and organize experience. This reflection is an integral part of reflective abstraction that consists of drawing properties for situations by paying conscious attention to the actions, interiorizing those actions into processes, encapsulating the processes into objects and finally organizing related actions, processes, objects into mental entities called schemas. In particular, the reflection on a schema with intention to transform it extends a learner's understanding by yielding an additional means of constructing an object (Cottrill et al., 1996). Thus, APOS theory accounts for the construction of objects from two different sources—encapsulation of processes and *reflections* upon schemas.

Another essential construct of APOS theory related to schemas is a theoretical description called a *genetic decomposition*. The fact that schemas are mental entities engenders one of the inherent difficulties of examining a learner's schema. In order to overcome this, although not in entirety, Dubinsky and others utilize genetic decompositions of concepts to characterize the linkages and representations within a concept. A genetic decomposition of a concept derives from three sources: psychological data, Piagetian ideas of concept formation, and understandings of the concept held by mathematicians (Dubinsky, 1991). The psychological data draw from observations of students in the midst of learning a concept. Piaget's ideas of successive refinements influence the construction and revision of the modeling genetic decomposition. These refinements engender from reflections on the learning experiments which guide the adjustments to the model thereby better accommodating new and relevant phenomena to produce a richer and more representative model. As the model requires less modification to account for the data generated from a teaching experiment, greater evidence exists that the model is descriptive. Lastly, mathematical descriptions of the concept are essential since the genetic decomposition must make sense from a mathematical perspective. However according to Dubinsky (1991), the genetic decomposition does not necessarily reflect how a particular mathematician would analyze the concept to formulate a method for teaching the concept. As a result, a genetic decomposition is an idealized characterization of the mathematically expected representations, linkages, objects, processes, and actions generally ascribed to the concept. In addition, the genetic decomposition provides a possible path for a learner's concept formation; however, it may not be representative of the path taken by all students (Dubinsky, 1991).

Asiala et al. (1996) assert "Our tentative understanding suggests that an individual's schema for a concept includes her or his version of the concept that is described by the genetic decomposition, as well as other concepts that are perceived to be linked to the concept in the context of problem situations" (p. 12). As a result, a learner's schema may or may not represent the whole or even a part of the genetic decomposition. This schema may lack essential elements or contain elements not considered mathematically connected to the concept. However, as Dubinsky (1991) pointed out, "It is not possible to observe directly any of a subject's schemas or their objects and processes. We can only infer them from our observation of individuals who may or may not bring them to bear on problems—situations in which the subject is seeking a solution or trying to understand a phenomenon. But these very acts or recognizing and solving problems, of asking new questions and creating new problems are the means (in our opinion, essentially the only means) by which a subject constructs new mathematical knowledge" (p. 103). As a result, attempting to uncover a learner's schema by presenting new tasks, observing the outward displays, and making inferences about the internal actions, processes, objects, and schemas from them appears futile. The futility arises from the possibility that the engagement with a task encourages the reorganization of thinking. However, consistent responses across multiple tasks indicate the learner assimilated a portion of a schema exemplified in a genetic decomposition.

4. Examining shared and distinctive elements

Pirie and Kieren's model of the growth of understanding and Dubinsky's APOS theory hold many elements that are isomorphic to one another and a few where they diverge. For instance, one can examine why Pirie and Kieren's model of the growth of mathematical understanding and Dubinsky's APOS theory satisfy the criteria associated with the status of a theory, compare the origins of the two theories, relate their organizational structures, examine the linkages to alternative theories of understanding, and discuss the implementation of the theories with respect to assessment and instructional practices. As a result, the following sections further identify these comparable qualities as well as the qualities which make the two theories distinct.

4.1. Why do these two frameworks satisfy the criteria for a theory? Prior to turning to the question of whether Pirie and Kieren's model and APOS theory truly satisfy criteria for a theory, a few terms need to be clarified. A *model* is a representation useful only to the degree in which it describes the linkages between represented objects and organizes a structure to help us understand those objects. In general, a model may systematically obscure particular features in an attempt to simplify relationships thereby leaving much of a situation unrepresented. Consequently, the model provides a working description of reality focused on the objects and their relationships without claiming absolute truth. Similarly, a *theory* does not claim absolute truth and its explanations focus on the level of mechanism. That is, it focuses on overarching meta-relationships rather than represented objects in a particular situation while still grappling with the working description of complex phenomena. Lastly, the term *theoretical framework* which has been used in this paper is a descriptive of either a model or theory which has not been scrutinized according to the criteria of Schoenfeld (2000).

So, what are Schoenfeld's criteria? According to Schoenfeld (1998), three criteria defined whether a theoretical framework could be classified as a theory: explanatory power, predictive power, and scope. These three criteria have been expanded in Schoenfeld (2000) to include: (a) descriptive power, (b) explanatory power, (c) scope, (d) predictive power, (e) rigor and specificity, (f) falsifiability, (g) replicability, and (h) multiple sources of evidence ("triangulation"). As each requirement is defined, a discussion will present how Pirie and Kieren's model of the growth of mathematical understanding and Dubinsky's APOS theory satisfy the criteria.

4.1.1. *Descriptive Power.* Descriptive power involves the capacity of the theoretical framework to capture the essential features under investigation in ways that permit faithful examination of the phenomena being described (Schoenfeld, 1998). Clearly, both Pirie and Kieren's model of the growth of understanding and Dubinsky's APOS theory satisfy this first criteria. Pirie and Kieren utilize interview transcripts and graphical images to characterize the movement of a student between levels of their model to identify the types of understandings utilized to answer various questions during the interview process. Similarly, Dubinsky's APOS theory uses both written and oral data collection techniques to form a descriptive document characterizing the student's "achieved" level of understanding. In this sense, both theories amply provide sufficient description for which a reader can interpret the theory and recognize the data correspondent to the conclusions. Both of these theories provide the observer of students' external actions or utterances with a means of collecting, organizing, and analyzing those observations. At the same time, they both are incomplete, leaving peripheral elements unexplained to characterize the focal issues.

4.1.2. *Explanatory Power.* Explanatory power refers to a framework's ability to explain mechanisms—descriptions of how and why things fit together and work. According to Schoenfeld (2000), the explanations provide the underlying reasons for why a student can or cannot perform a particular task. In essence, a theory must contain precise, descriptive terms which indicate the important objects of the theory, their interrelationships, and the reasons particular things are possible or not. Dubinsky and McDonald (2001) argue that the explanatory power of APOS theory resides in the theory's ability to point to particular mental constructions of actions, processes, objects, and/or schemas that one student appears to have made whereas the other has not. In a similar fashion, Pirie and Kieren's theory allows them to explain differences in student performance based upon the level of understanding correspondent to prior concepts. For instance, in Pirie and Kieren (1992a), they differentiate between the performance of students based upon their ability to work with "thirds" based upon the students' external actions and utterances providing indication of differences in the layer of operation. In particular, their theory points to developed images, noticed relationships between images, formalized descriptions of those relationships, etc. as explanations for variances in student performance.

4.1.3. *Scope.* Scope involves the range of phenomena attended to by the theory. Essentially, a comprehensive theory must apply to a broad range of phenomena rather than a localized concept. APOS theory has been widely employed by Dubinsky and members of his RUMEC team across topics such as: functions, abstract algebra (binary operations, groups, subgroups, cosets, normality, quotient groups), discrete mathematics (induction, permutations, symmetries, existential and universal quantifiers), calculus (limits, chain rule, derivative, infinite sequences), statistics (mean, standard deviation, central limit theorem), elementary number theory (place value in base n, divisibility, multiples, converting between bases), and fractions. The number of concepts mentioned clearly points to the applicability of APOS theory to a broad range of phenomena typically linked to undergraduate mathematics, although a few of the topics have been investigated with younger children (Dubinsky & McDonald, 2001). In contrast, Pirie and Kieren's model of the growth of understanding has generally focused on the development of understandings in younger children. Consequently, the range of phenomena for which Pirie and Kieren's theory has been applied is smaller than that of APOS theory. Concepts such as fractions, quadratic functions, and other middle school content have been the focus of their investigations with a study of geometry learning in Pirie and Kieren (1991a). However, others have employed Pirie and Kieren's model of understanding with respect to calculus (Meel, 1995), abstract algebra knowledge (Grinevitch, in preparation), and teacher interventions (Towers, Martin, & Pirie, 2000) thereby raising the possibility that Piric and Kieren's model may be applied to a broader range of phenomena than middle school mathematical concepts.

4.1.4. *Predictive Power.* Predictive power is not at the level of those made in the physical sciences but rather refers to the ability of the theory to provide reasonable predictions as to the observed actions and utterances based upon prior information. In essence, predictive power permits the researcher to anticipate responses, based upon prior knowledge, before the participant actually responds. Pirie and Kieren's model of the growth of understanding draws upon the weight of prior information elicited from a student in order to identify the images and structures built by a student. Using this knowledge, Pirie and Kieren's theory both suggests potential actions and utterances of a student with respect to a new task based upon their experience with prior tasks. For instance, in Pirie and Kieren (1992a), they state the following about a student: "Sandy, a gifted eight year old, who had shown formalized understanding with respect to 'half fractions,' appeared to apply a formal method to generate new fractions" (p. 515). As students utilized their knowledge of "half fractions" and attempted to extrapolate this to "third fractions," it was clear that Pirie and Kieren's theory permitted the teacher to anticipate the potential images developed and how the students would interact with those newly-developed images. Further instances of making testable predictions of student responses were found in Towers, Martin, and Pirie (2000) with respect to teacher interventions. In a similar fashion, APOS theory permits the development of predictions. According to Dubinsky and McDonald (2001), APOS theory provides the opportunity to make testable predictions "that if a particular collection of actions, processes, objects and schemas are constructed in a certain manner by a student, then this individual will likely be successful using certain mathematical concepts and in certain problem situations" (p. 4). The *genetic decompositions* of concepts employed by APOS theory both provide descriptive information as well as a means of generating hypotheses about how the learning takes place and what elements interact with the development of an individual's understanding of a particular concept.

4.1.5. *Rigor and specificity.* Rigor and specificity refer to the ability of a theory to clearly identify the elements inherent to the theory and the mechanisms which connect them. Explicitly, rigor and specificity are concerned with the terms and relationships of theory being well-defined, that is, if you were interviewing a student could you easily detect that they were operating at a particular layer in Pirie and Kieren's model or with a particular conception in APOS theory. Earlier in this paper the elements and constructs of both Pirie and Kieren's model and Dubinsky's APOS theory were defined. These characterizations not only provided the descriptions of the elements and their linkages but identified their relationships to other related perspectives. For examples of student data and a means of interpreting the conversations, examine the Pirie and Kieren (1994b) description of Teresa or Julie working with fractions, the Pirie and Kieren (1992) description of Sandy, the section 7.1.4 description of Ada in Brown et al. (1997) who was in flux between action and process conception of a group, or the chapter 7 characterization in Cottrill (1999) describing students, such as Tim, Al, Ray, Peg, and Jack, from the perspective of operating at either the Intra, Inter, or Trans level.

4.1.6. *Falsifiability.* Falsifiability is needed by any theory. In practice, falsifiability means that the predictions made and elements defined by theory must hold up to empirical examination. In fact, both Pirie and Kieren's model and Dubinsky's APOS theory do not claim their frameworks are truth but serve as a descriptive language needing scrutinization and testing. For instance, Dubinsky and McDonald (2001) state:

> We do not think that a theory of learning is a statement of truth and although it may not be an approximation to what is really happening when an individual tries to learn one or another concept in mathematics, this is not our focus. Rather we concentrate on how a theory of learning mathematics can help us understand the learning process by providing explanations of phenomena that we can observe in students who are trying to construct their understandings of mathematical concepts and by suggesting directions for pedagogy that can help in this learning process. (p. 1)

Similarly, Pirie (1988) states:

> In all actuality, we can never fully comprehend "understanding" itself. As Piaget (1980) claims for all knowledge, with each step that we take forward in order to bring us nearer to our goal, the goal itself recedes and the successive models that we create can be no more than approximations, that can never reach the goal, which will always continue to posses undiscovered properties. What we can, however, do is attempt to categorize, partition and elaborate component facets of understanding in such a way as to give ourselves deeper insights into the thinking of children. (p. 2)

Consequently, neither Pirie and Kieren's model or APOS theory claim to be truth but rather claim to provide a language for the mechanisms necessary to describe development in a variety of settings.

From this perspective, both Pirie and Kieren and their colleagues and Dubinsky and his RUMEC colleagues have continued to empirically test the applicability of their respective frameworks to a variety of mathematical concepts and under a variety of settings. As a new concept is explored or a new setting is examined, the researchers continually refine and adapt their frameworks to better explain the phenomena thereby transcending prior descriptions while maintaining compatibility with prior assertions. In this manner, these frameworks provide dynamic lenses for viewing situations and the interactions contained within those situations. Consequently, any challenges to their veracity must be focused on a framework's inner consistency and its ability to interpret situations under those constraints.

4.1.7. *Replicability.* Replicability is closely tied to rigor and specificity. According to Schoenfeld (2000) "There are two related sets of issues: (1) Will the same thing happen if the circumstances are repeated? (2) Will others, once appropriately trained, see the same thing in the data" (p. 648)? In particular, replicability is concerned with a framework's ability to describe similar behaviors in similar manners as well as different groups of researchers trained in a particular framework seeing and describing similar events with similar language. That is, the framework must be defined well enough that others can look at the same data and reach the same conclusions.

Generally, replicability is concerned with the ability of a researcher to examine the procedures and constructs, implement a study using those procedures and constructs, and interpret the data in similar fashions. Discussion of the application of both Pirie and Kieren's model and Dubinsky's APOS theory to the analysis of particular mathematical concepts clearly indicate their respective data collection methods, employed theoretical structures, mode of analysis, and general results or parsed transcripts. By being able to review the language and work of the participants in these studies, one can gain a vision of the interaction of the theories and the data along with a confidence of the repeatability of such methods to other participants and situations. However, the added condition of repeatability should not be construed with certainty. Pirie and Kieren's model and Dubinsky's APOS theory provide a range of possible outcomes assuming a student interprets questions in manners anticipated by the researchers (see section 4.5.2 on *Instructional practices* for further discussion).

4.1.8. *Triangulation.* The last criterion identified by Schoenfeld (2000) was that of triangulation. That is, a theory must not be developed based upon a limited set of experiences but rather must utilize information collected in a variety of settings (classroom interviews, individuals working on tasks, pairs working together, etc.). In doing so, the theory can be examined for internal consistencies as well as oversensitivity to local factors not influential to the general case. In addition, triangulation refers to the theory's ability to utilize multiple sources of information when analyzing a particular concept as well as being informed from a broad range of phenomena. As previously argued, both Pirie and Kieren and Dubinsky fine-tuned their theories over a wide spectrum of phenomena and in a variety of settings by gathering evidence on students from a variety of sources. With respect to the use of internal triangulation, Pirie and Kieren's model focuses on interview data collected on an individual basis, as part of pairs or groups, and as part of a class. Interpretations concerning a student's placement with respect to the theory also come from the learner's interaction with multiple tasks designed to help the researcher observe the changes in mathematical knowing. On the other hand, Dubinsky's APOS theory utilizes both written work and interview data to coalesce a vision of the actions, processes, objects and/or schemas developed by a student. Using this two-pronged approach, data collection methods used by APOS researchers afford the researcher with a global overview from written work and then the ability to focus on particular "bits" of knowledge during the interviews.

This discussion has provided a means of validating the assertion that Pirie and Kieren's model of the growth of understanding and Dubinsky's APOS theory satisfy the criteria set forth by Schoenfeld (2000). The rest of this section will further identify the qualities which connect the two theories as well as those qualities which make them distinct.

4.2. Origins of the theories. Both Pirie and Kieren's model of the growth of understanding and Dubinsky's APOS theory arise from a constructivist point of view. Pirie and Kieren's model arose from Glasersfeld's perception that understanding is a continual process of organizing one's knowledge structures whereas Piaget's discussions of reflective abstraction drove much of the construction of the APOS theoretical framework. According to each theory, the learner actively participates in the enterprise of constructing understanding from the elements and situations provided in mathematical problems. This construction, according to both Pirie and Kieren's model and APOS theory, is a process where learners continually engage in the act of organizing their own knowledge structures (Ayers et al., 1988; Pirie & Kieren, 1991a). As a result, both theories describe understanding as a never-ending process. In addition to this bond, the ties between the theories do not end with their constructivist roots. Both theories developed out of a tradition of significant observation and interaction with students engaged with particular mathematical content. It is with this respect that the two theories are both similar and different.

The primary origin of Pirie and Kieren's model rests in the observation of the teaching and learning of middle and high school level mathematics such as fractions and quadratic functions (Pirie & Kieren, 1989, 1991a). Such observations occurred as part of whole class "teaching experiments" with Tom Kieren acting as the teacher and students were involved in a constructivist environment (Pirie & Kieren, 1994a). Students similar to "Sandy" (Pirie & Kieren, 1992b) were selected from the larger learning environment for individualized interview sessions. Concurrent with the interviews, class activities were audio or videotaped while research assistants tracked the work of subgroups of students, augmenting transcripts with field notes, and collected student work (Pirie & Kieren, 1992a). These various artifacts were studied for the presence of activities which allowed Pirie and Kieren to test, elaborate, and develop constructs and properties of their theory.

Pirie and Kieren (1992b) describe an overall learning environment in which students are engaged for a couple of weeks with the area of fractions using the "half family" and paper-folding explorations. An excerpt from an interview with Sandy then is used to illustrate the various interviewer questions and Sandy's responses. One element to be noticed is that the interviewer's questions were not fixed but rather reacted to Sandy's responses while posing new questions, elaborating on old ones, or achieving understanding of the student's responses. Using the transcript, the researchers proceeded to produce a fine-grained map of the growth of Sandy's understanding as it is observed during the interview. The phrase "as it is observed" is of particular importance since Pirie and Kieren (1994) state "we make no claims as to what might have gone on 'in the students' heads'" (p. 182). As a result, all inferences about what images were tapped to answer questions, how Sandy returned to previous ways of knowing in order to extend understandings, and the stability of the interviewer's interpretations of Sandy's responses (for further discussion, see Pirie and Kieren, 1992b, pp. 249–255) were mediated by the learner's reactions to the questions and triangulated against the learner's other responses.

Integral to the overall process of theory validation is a reflection on the teaching components utilized to spur students to actively engage with the content and personally explore the interconnections of content. However, the results of designed teacher interventions, even those with a specific intent, depend on the actions and interpretations of the students (Pirie & Kieren, 1992a). In fact, students may interpret a teacher's comments in unintended ways causing unexpected responses.

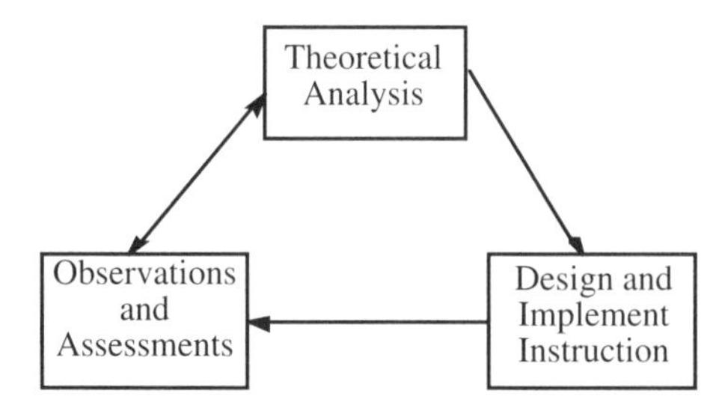

FIGURE 6. The APOS Cyclical Research Framework

Consequently, the mere examination of what a student does in the face of a mathematical task is not sufficient (Pirie & Kieren, 1992a). For example, questions posed by the interviewer or teacher need to be analyzed with respect to the student's response rather than the intended effect. Additionally, the interviewer or teacher must recognize that students can understand things at many levels and therefore investigators should expect students to respond at various levels at various points throughout a class or interview.

In contrast to Pirie and Kieren's focus on the growth of student understanding of middle school mathematics, APOS theory developed out of the observation of students working with undergraduate abstract algebra, functions and calculus, discrete mathematics, and elementary number theory (Dubinsky & McDonald, 2001). This development, according to Asiala et al. (1996), continues to involve a cycle of research (see Figure 6) beginning with a theoretical analysis of the elements perceived as necessary for the development of conceptual understanding, the design of instructional activities to compel students to make the requisite constructions, the observation and assessment of student constructions, and the reflections on the theory's ability to capture the nuances of those constructions. The particular composition of this research cycle involves the blend of (a) observing students in the process of learning mathematical concepts in order to identify the development of conceptual structures, (b) analyzing the data in light of the underlying APOS theory to generate a genetic decomposition indicative of one possible way a student might construct the concept, (c) designing instructional activities to help move students to make those constructions and thereby make the particular reflective abstractions indicated by the genetic decomposition, and (d) repeating the process after reflection on the genetic decomposition and the instructional treatment (Dubinsky, 1991). This research cycle continues to spiral upon itself, combined with explicit reexamination of the theoretical analysis' ability to describe the various nuances uncovered during observation and assessment, until stability is reached between the theoretical analysis and its ability to explicate what it means to learn a particular concept and how the suggested instructional treatments help students succeed in that learning (Dubinsky, 1994).

Integral to such a research cycle is the concerted examination of student learning in classroom situations where in many cases the *ACE Teaching Cycle* (described later in this paper) was employed. For example, Asiala et al. (1996) describes a data collection focused on student responses to written instruments and individualized interviews. The interviews were audio-taped, transcribed, and correlated with ancillary documentation produced by the student (Asiala et al., 1996). These data were then used to "evaluate the usefulness of our models of cognition by asking whether the mental constructions proposed in the initial analyses appeared to be made by the students. Second, ... [to] evaluate the mathematical performance of the students on tasks related to the concepts" (Asiala et al., 1998, p. 15). In effect, a concept's genetic decomposition was analyzed in light of the various responses to the instruments and interviews in order to determine the efficacy of the model to detect and explain differences in understanding by pointing to specific mental constructions of actions, processes, objects and/or schemas held by the learner.

For instance, the APOS theory-related research focuses on the ability of the genetic decomposition to describe potential mental constructions as the learner moves through various stages of cognitive development. In Cottrill et al. (1996), the researchers describe the evolution of a genetic decomposition of the concept of limit. The resulting decomposition that resulted was:

- The action of evaluating the function f at a few points, each successive point closer to a than was the previous point.
- Interiorization of the action of Step 1 to a single process in which $f(x)$ approaches L as x approaches a.
- Encapsulate the process of Step 2 so that, for example, in talking about combination properties of limits, the limit process becomes an object to which actions (e.g., determine if a certain property holds) can be applied.
- Reconstruct the process of Step 2 in terms of intervals and inequalities. This is done by introducing numerical estimates of the closeness of approach, in symbols, $0 < |x - a| < \delta$ and $|f(x) - L| < \epsilon$.
- Apply a quantification scheme to connect the reconstructed process of the previous step to obtain the formal definition of limit ...
- A completed ϵ–δ conception applied to specific situations. (Cottrill et al., 1996, pp. 174–175)

The researchers then implemented a nearly seven-week series of instructional activities engaging calculus students with tasks involving both computer and group work. These activities were designed to help the students make the mental constructions defined in the above genetic decomposition and included computer investigations into approximations, graphical representations of the limit concept, computer-aided constructions of a value approaching a limit, computer-based construction of the limit concept, and ϵ–δ windows.

Drawing from interviews of selected students, the genetic decomposition was revised to better reflect the constructions being observed. One should note, however, that similar to Pirie and Kieren's belief concerning the intractability of observing the internal cognitions of a learner, the founders of APOS theory agree with Glasersfeld's radical constructivist[8] position, according to Asiala et al. (1996), and states that "it is impossible for one individual to really know what is going on in the mind of another individual" (p. 29). As a result, the revised genetic decomposition leads to continued observation and revisitation to the overall theory and either its

[8]Glasersfeld disagreed with the Piagetian assertion that an individual's concepts of reality could be mirrored and instead supported the idea that reality lies beyond an observer's scope of experiential justification. Consequently, he believed that even though one can choose well-defined tasks or experiences, the constructive process achieves no prescribed ends and that it is untenable that there exists some perfect internal representation or match against which one can test a learner's understandings, i.e., constructed reality.

revision and refinement or in rare cases, an enhancement of the theory to further explicate observed phenomena not describable in terms of the original theory. In particular, Cottrill et al. (1996), noted that an even more preliminary step to the genetic decomposition involving the action of evaluating f at a single point near or equal to a was needed. Also, Steps 2 and 3 of the original genetic decomposition were revised to allow for the possibility of multiple processes leading to the construction of a coordinated schema which could then be acted upon and eventually encapsulated. Using this revised genetic decomposition, Cottrill et al. (1996) were able to reconcile their observations drawn from the interview data and make attributions for differences in student performance as well as provide pedagogical suggestions for increased attention to domain and range processes and how to use the function to coordinate them.

Clearly, Pirie and Kieren's model and Dubinsky's APOS theory are grounded in a great deal of observation and teacher interactions with students actually doing mathematics. They both grew from similar constructivist roots and this underlying foundation has resulted in considerable commonalities between the theoretical frameworks. For example, the search for a means of describing mathematical understanding, and ability to detect fine-grained, observable differences between students. However, the development of APOS theory from investigations into undergraduate mathematical content and APOS researchers' usage of genetic decompositions indicates an emphasis on the *mathematical* portion of mathematical understanding. In contrast, Pirie and Kieren focused on the *understanding* element by relegating the mathematical content to background. Even though the foci are different, both theories are sensitive to the mathematical actions of students and integrate on-going observations of student actions in context to have the researchers reevaluate and revise their theories thereby better describing the learner's constructions.

4.3. Organization and mechanisms of the theories. Pirie and Kieren organized their model into eight layers—Primitive Knowing, Image Making, Image Having, Property Noticing, Formalizing, Observing, Structuring, and Inventizing—with each layer delineating a qualitative change in growth of a learner's understanding. In addition, the model contains a construct referred to as the "don't need boundaries." When a learner passes a "don't need" boundary, the learner exhibits the capability "to work with notions that are no longer obviously tied to previous forms of understanding, but these previous forms are embedded in the new layer of understanding and readily accessible if needed" (Pirie & Kieren, 1994b, p. 172). As a result, the layer closest to an inner "don't need" boundary indicates qualitatively different understanding in comparison to the previous layer whereas the layer closest to an outer "don't need" boundary involves taking those new understandings and coordinating them into richer relationships. The sections between two successive "don't need" boundaries behave as units which first involves the construction of a new way of conceiving and then the organization of these new conceptions into a form capable of being transformed into a higher layer of conception.

The three "don't need" boundaries partition Pirie and Kieren's eight layered model of understanding into four units that have strong connections to elements of Dubinsky's APOS theory. In particular, the first "don't need" boundary occurs between the *image making* and the *image having* layers since a learner gains an image of a concept and does not need external actions or the specific instances of image making (Pirie & Kieren, 1994b). This description is similar to the differences between the action and process conceptions described by Dubinsky et al. (1994) where a *process* conception indicates the learner internalized the external actions thereby allowing the coordination and reversal of the process to form new mental processes. This first "don't need" boundary in Pirie and Kieren's model of understanding separates the external actions and the newly developing internal processes that the learner built upon these actions.

The next "don't need" boundary in Pirie and Kieren's model of understanding comes between the *property noticing* and *formalizing* layers. In this instance, the learner moves from having images of a concept to coalescing these into a formal mathematical idea (Pirie & Kieren, 1994b). In particular, overcoming this boundary indicates the learner's diminished reliance on images and permits the learner to conceive of the mathematical ideas as a class-like mental object (Pirie & Kieren, 1992). In APOS theory, movement from a *process* conception to an *object* conception occurs "[w]hen an individual reflects on operations applied to a particular process, becomes aware of the process as a totality, realizes that transformations (whether they be actions or processes) can act on it, and is able to actually construct such transformations, then he or she is thinking of the process as an object. In this case, we say that the process has been encapsulated to an object" (Asiala et al., 1996, p. 11). Thus, advancement to an object conception involves coalescing a concept's processes into an object no longer requiring the underlying processes but retaining them in the sense of Sfard's operational/structural dichotomy.

The last "don't need" boundary described by Pirie and Kieren resides between the *observing* and *structuring* layers of their model. Transcending this boundary, according to Pirie and Kieren (1994b), indicates that the learner has developed a mathematical structure and does not require the meaning brought to it by inner layers of understanding. Similarly, the movement between an *object* conception and a *schema* conception involves the learner in constructing a coherent structure of actions, processes, objects and subordinate schemas that exist in the mind of the learner. This structure permits the learner to deal with problem situations by allowing the schema to be unpacked and work with the underlying objects and processes which it encompasses (Dubinsky et al., 1994). Thus, the "don't need" boundaries of Pirie and Kieren's model of understanding partition their model into four overarching units similar to the four levels of Dubinsky's APOS theory. The primitive knowing and image making layers correspond to the action conception, image having and property noticing layers corresponds to the processes conception, formalizing and observing layers necessitate an object conception, and lastly structuring and inventizing layers organize a structure similar to a schema conception.

In addition, both theories contain a comparable mechanism for the extension of understanding. Pirie and Kieren's model of understanding provides folding back as a generative mechanism that corresponds to a learner's return to an inner layer of understanding to augment, restructure, or expand the inner layer's inadequate understandings (Pirie & Kieren, 1994b). In the case of folding back, this adjustment to the inner layer understandings receives guidance from the experiences and understandings constructed as part of the outer layers. *Inadequate* can mean two different things: incomplete or insufficient. *Incomplete* implies that additional elements must be added to a lower layer of understanding to support the development of understanding at a higher layer. This type of folding back tends to occur when a learner has not fully reached the formalizing layer and must restructure the lower layers to incorporate new images necessary to build a formalized understanding. *Insufficient* suggests the scope of understanding at a lower layer needs expanding to inform the extension of understanding to a new set of circumstances. For instance, a student may have a formalized understanding of adding pairs of fraction using a personally generated cross multiplication trick but when faced with having to add three fractions, the student may fold back to image making to create a method to accomplish this (Pirie & Kieren, 1992). In this case, such activity consists of expanding the scope of the understanding to handle addition of three fractions but this expansion receives direction from the formalized understandings already present thereby suggesting possible viable strategies for the expansion.

This latter description corresponds to Dubinsky's portrayal of *de-encapsulation* which is a return "*to the process which was encapsulated in order to construct the object in the first place*" (Dubinsky, 1994, p. 271). In particular, de-encapsulation permits the learner to return to the process origins of objects, coordinate those processes to form new processes that can themselves be encapsulated into new objects. As a result, de-encapsulation serves the generative role of returning to previous ways of thinking to provide a foundation for the extension of previous conceptualizations. For instance, to understand addition of functions, Dubinsky (1991) asserts that the learner must de-encapsulate the function objects to their process-oriented roots, those processes must be coordinated through point-wise addition, and then the resulting process encapsulated to form a new object. As a result, the de-encapsulation mechanism of Dubinsky's APOS theory permits the extension of understanding by expanding the scope of previously held understandings similar to the folding back mechanism of Pirie and Kieren's model of understanding.

4.4. Linkages to other theoretical frameworks. Pirie and Kieren's model of the growth of understanding, Dubinsky's APOS theory, and the other contemporary theoretical frameworks[9] of understanding all form different but compatible lenses for the development of understanding. In particular, each framework has been developed to provide researchers and teachers means of observing understanding as an ongoing process in which interpretation is predicated upon the student's personal knowledge structures, the social dynamics of the learning situation in which the learning occurs, and the constraints on the display of those understandings due to the nature of the environment. Given the content and setting in which researchers attuned their particular framework, each focused on different components and uncovered specific connections between those components. This yields a collection of theoretical frameworks where each provides a portrait of students engaged in the development of understanding; but as is the case with any portrait, certain features of that development are highlighted and others are de-emphasized. In this next section, the goal will be to further explicate the relationships, if any, that Pirie and Kieren's model of the growth of understanding and Dubinsky's APOS theory have with the theoretical frameworks associated with *cognitive* or *epistemological* obstacles, *concept definition* and *concept image*, *multiple representations*, and *operational* and *structural* entities.

[9]The continued use of "theoretical framework" in reference to the other recent work on understanding is not meant to convey that they do not satisfy the requirements identified by Schoenfeld (2000). It is quite conceivable that some if not all do satisfy these requirements but this paper has not focused itself on attempting to argue this status for them.

4.4.1. *Errors, misconceptions, and epistemological obstacles.* Researchers attribute errors and misconceptions to extrapolations based on insufficient examples, ambiguous examples, missing linkages, or even partial linkages between relevant pieces of knowledge such as conceptual and procedural knowledge (Hatano, 1988; Hiebert & Carpenter, 1992; Hiebert & Wearne, 1986; Mansfield & Happs, 1992; Silver, 1986; Tall, 1989; VanLehn, 1986). According to Ferrini-Mundy and Gaudard (1992), these errors act as windows for observing the inner workings of a learner's mind and correspond to elements in the theoretical framework of epistemological obstacles previously described in this paper. Both Pirie and Kieren's model of understanding and Dubinsky's APOS theory contain components that account for the development and surmounting of misconceptions and epistemological obstacles.

In particular, Pirie and Kieren account for the possible development of misconceptions when discussing movements within their model. Such movement requires the construction of additional connections eventually reaching the structured layer where the network has become a stable entity. In the process of moving from the image having layer to the property noticing layer and then the formalizing layer, the learner proceeds through a process of internalizing images, recognizing properties associated with groups of images and integrating the various recognized properties into formalized mental objects that have no internal contradictions (Pirie & Kieren, 1991a). A learner moving from image having to formalizing does not have to overcome a misconception in making such a transition since the potentially troublesome images may not have been constructed. However, engagement in property noticing on a limited domain of images can result in the extrapolation of connections from the evoked images which may be incomplete in the domain of all images associated with a topic. This construction, although not necessarily encountered, can result in the development of a troublesome image considered a misconception. In the case that such a misconception arises, the movement from property noticing to formalizing signifies the learner has overcome some of the property-based misconceptions by recognizing inconsistencies with established connections, developing correct connections, forging elaborate networks of connections, and finally constructing mental objects.

Many of the same mechanisms for developing and overcoming misconceptions identified in the discussion of Pirie and Kieren's model also exist in APOS theory. For instance, Dubinsky et al. (1994) mention that "the general theory of Piaget (1975) includes the idea that concepts are constructed at a layer that is adequate for dealing with a learner's current mathematical environment, but that when new phenomena are confronted, it is necessary to reconstruct concepts on a higher layer. Thus, if the construction is delayed, a student's conception of some mathematical notion may be adequate at one layer, but erroneous at another" (p. 295). A misconception, from Dubinsky's perspective, is an understanding that has not reconciled itself with a broader context but remains reasonable in a narrower scope. Thus, Dubinsky (1991) uses a "lack of understanding" to mean that the student has not constructed an appropriate schema for the examined concept. However, overcoming an epistemological obstacle equates with a learner interacting with a disequilibrating environment, reflecting on held understandings, and reorganizing those understandings to integrate the discrepant elements of the perplexing situation.

4.4.2. *Connections to concept images.* As already mentioned, both Pirie and Kieren's model of understanding and Dubinsky's APOS theory have characteristics associated with *concept image.* In particular, Pirie and Kieren (1992a), although not directly referring to the term *concept image*, state that "young children use their own created mental objects, some of which do have a figural quality but others of which are abstract, to build their mathematical knowledge" (p. 505). This figural quality particularly arises at the image having and property noticing layers although existent at other layers. In addition, such a description echoes the definition of concept image as a collection of mental pictures, representations, and related properties ascribed to a concept by a learner. In fact, Pirie and Kieren (1994b) commented that their use of the word "image" in labeling two of their layers was due to "evidence at these levels is frequently based upon pictorial representation" (p. 166); however, understanding at those levels are not restricted to only this mode of expression but can include mental imagery which is defined by the reasonably established concept of mental objects. On the other hand, the word "idea" could have been used but it was thought that "image" was less ambiguous and carried many of the connotations sought by the theory.

In addition, Pirie and Kieren (1992b) draw connections to the work by Harel and Tall (1991) concerning generalization. As Harel and Tall (1991) detail different student mental constructions, they tacitly use phrases such as "the subject reconstructs an existing scheme" and Pirie and Kieren (1992b) connect this explicitly to one of the essential features of their theory, namely folding back. Also of interest is a belief that disjunctive generalization integrates with Pirie and Kieren's (1992b) general idea of disconnected understanding as well as that generic abstraction "accords well with our theory that understanding grows by noticing properties of one's images and by making observations on one's formalizing" (p. 245). Consequently, the images and processes connected within the theoretical framework of concept images have strong ties to elements of Pirie and Kieren's model.

The theoretical framework of concept definition/image, according to Dubinsky (1991), explains why understanding fails to develop whereas reflective abstraction describes the essential elements for the development of understanding. However, Asiala et al. (1996) assert that the notion of concept image provides explanations beyond why understanding did not develop. They consider the concept image to correspond with Piaget's schemata which is closely linked to APOS theory's use of schema. According to Dubinsky (1996), "A schema is a more or less coherent collection of objects and processes. A subject's tendency to invoke a schema in order to understand, deal with, organize, or make sense out of a perceived problem situation is his or her knowledge of an individual concept in mathematics. Thus an individual will have a vast array of schemas" (p. 102). As a result, a concept image and a schema serve the same purpose of organizing and structuring held understandings concerning a concept. Additionally, the last two sentences quoted from Dubinsky resonate with Tall and Vinner's description of evoked concept images which involve the portions of a concept image activated by particular stimuli. In fact, Tall and Vinner (1981) assert that "different stimuli can activate different parts of the concept image, developing them in a way which need not make a coherent whole" (p. 152). Thus, a strong connection between concept images and schemas exists.

4.4.3. *Connections to multiple representations and operational and structural conceptions.* Pirie and Kieren do not make any specific reference to the ideas of multiple representations as described by Kaput. However, a discussion of Sfard's duality of operational and structural conceptions occurs in some papers by Pirie and Kieren. For instance, Pirie (1988) in a paper briefly surveying several interpretations of understanding, i.e., instrumental, rational, intuitive, constructed, formalized, etc. mentioned that according to Sfard, an operational conception is the first step in the acquisition of a new mathematical idea. This assertion accounted for the linkage of an initial operative notion of a concept with constructing a beginning understanding. A broader discussion appears in Pirie and Kieren (1992b), where they identified Sfard's work as "interested in understanding at the level of algorithmic thinking" (p. 244). This characterization indicates a belief that Sfard contends the acquisition of mathematical ideas comes from initially operationally derived notions which in turn coalesce through interiorization, condensation, and reification into a structural view. The linkage of Sfard's work to Pirie and Kieren's model of understanding focuses on Sfard's concern for the formalizing activity in mathematical understanding. Sfard (1991) implies that the reification process, i.e., objectifying an operational notion, involves a difficult change on the part of the learner. It is Pirie and Kieren's belief that their theory provides description of several actions and mechanisms that are integral to this change (Pirie & Kieren, 1992b). However, Pirie and Kieren's (1992) belief in the non-linearity of the growth of understanding separates them from Sfard.

Similar to Pirie and Kieren's model of understanding, Dubinsky's APOS theory draws no explicit connections to Kaput's theoretical framework of multiple representations. However, Dubinsky (1994) mentions that he approaches the idea of multiple representations in a different manner than other researchers. Rather than using tools provided in some computer environment, Dubinsky devises investigations where students engage with the computer in developing their own tools for comparing different representations of a concept. Additionally, Dubinsky's APOS theory connects with Sfard's operational and structural duality (Dubinsky et al., 1994). Sfard and Linchevski (1994) expound upon this linkage when they ascribe Dubinsky's APOS theory, as described in Dubinsky (1991), as providing a closely related model to their process-object duality of mathematical concepts. In particular, they assert that the main connection between these two frameworks is the Piagetian description of reflective abstraction.

In addition, the term "reification" and Dubinsky's use of "encapsulation" have similar meanings since both arise from Piaget's description of the operational origins of mathematical notions (Sfard, 1994). In particular, Sfard (1991) states "The pioneering work in this field has been done by Piaget, who wrote in his book on genetic epistemology (1970, p. 16): 'the [mathematical] abstraction is drawn not from the object that is acted upon, but from the action itself. It seems to me that this is the basis of logical and mathematical abstraction'" (p. 17). Consequently, Sfard contends that both Dubinsky's work and her own work further elaborated Piaget's original ideas and that her characterization of a dichotomy between operational and structural conceptions provides a broader conjecture for the duality of mathematical thinking.

4.4.4. *Linkages of Pirie and Kieren's model of understanding to a refinement of APOS theory.* The refinement of the description of schema development provided by Clark et al. (1997) reveals an interesting connection to some layers of Pirie and Kieren's model of understanding. In particular, Clark et al. (1997) identified three stages in the development of a concept: *Intra*, *Inter*, and *Trans* previously described in Section 3.1. At the *Intra* stage, a learner has a collection of rules but has not recognized relationships between them. This is similar to the image having layer of Pirie and Kieren's model of understanding which is described as a level where images of a concept have been built from single activities but these images, although isolated, form a foundation for mathematical knowing (Pirie & Kieren, 1991a). The *Inter* stage occurs when the learner can organize the various cases encountered and recognize the relationships between these cases (Clark et al., 1997). This stage is isomorphic to the property noticing layer that indicates the learner can examine held images for specific and relevant properties to note their distinctions, combinations and connections and predict their actualization (Pirie & Kieren, 1991a). Attainment of the *Trans* stage indicates the learner has constructed an underlying structure for a particular concept. It is at this point that "The elements in the schema must move from being described essentially by a list to being described by a single rule" (Clark et al., 1997, p. 360). This description is similar to the formalizing layer which involves the learner in conscious contemplation of the noticed properties to abstract common qualities from them thereby possibly constructing a full mathematical definition to accompany the new class-like mental object (Pirie & Kieren, 1991a). Thus, each of the stages in this refinement of APOS theory reasonably correlate with the three layers of Pirie and Kieren's model of understanding: image having, property noticing and formalizing.

4.5. Assessment and instructional implementations. Both Pirie and Kieren's model of understanding and Dubinsky's APOS theory gather information on student understanding and question how to use this data to adjust instructional practices to promote the growth of understanding. The next two sections will attempt to provide information about the implications these theories have with respect to assessment and their ramifications for instruction.

4.5.1. *Assessment of understanding.* Pirie and Kieren (1989, 1992b), in concert with the opinions expressed by Pirie and Schwarzenberger (1988) as well as Simon (1993), recognize the utility of interviews to track a student's movements through the layers of understanding. In particular, Pirie and Kieren believe a written instrument, especially a multiple-choice exam, does not completely expose student understandings since they posit student understandings can only be inferred and not measured. They contend that written instruments are less facile to delve into the growth of student understanding since written instruments in their view provide static captions of student external demonstrations. As a result, they consider interviews as the primary means of uncovering a learner's changing understandings. These interviews permit one to make illations about a learner's awareness of relationships between concepts, ability to adapt procedures to novel situations, possession of generic examples, and fluency with language and symbolism.

Drawing inferences about movements between layers of understanding can only ensue after the learner performs actions and verbalizations (Kieren, 1990; Pirie &

Kieren, 1990). These form-oriented actions[10] and verbalizations demonstrate to an external agent the learner's operation at a level of understanding. Without such demonstration, the external agent might think the understanding has not been achieved. Integral to the establishment of a learner's level of understanding are situations that require the learner to act and verbalize in reference to unique and complex situations. These circumstances help exhibit a learner's present level of understanding and promote the development of understanding when instigated in an interview situation. The interview questions act as teaching tools and become a useful diagnostic instrument since they will either provoke understanding to an outer layer, invoke folding back, or validate the learner's understanding. Even though an investigator may wish to categorize questions, categorization can only occur after the hearer responds to a question demonstrating the effect of the question on the learner's changing understanding. In particular, a question which is provocative points a learner to an outer layer enabling continued development of understanding. In-vocative questions cause a learner to fold back to an earlier layer to continue progressing. Validating questions provide evidence that a learner is operating at a specific layer of understanding by encouraging the learner to demonstrate, either in verbal or symbolic forms, current mathematical actions (Pirie & Kieren, 1991a).

Thus, Pirie and Kieren's use of pro-vocative, in-vocative, and validating questions coincides with Glasersfeld's assertion that even though one might be "studying the construction of mathematical reality by individuals with in the space of their experience. In this construction, although there may be well-defined tasks or spaces for experience, there are no prescribed ends toward which this construction strives" (Steffe & Kieren, 1994, p. 721). As a result, Pirie and Kieren (1990, 1991a) take the position that one cannot give a student understanding rather the student will create her or his own understanding. In addition, Pirie and Kieren (1994b) state "Each person's mathematical understanding is unique. Indeed, since we believe that all knowledge is personally constructed and organized, students in any environment will construct understanding in some form" (p. 526). As a result, the interview process becomes a teaching tool and an opportunity to map the external evidences exhibited by a learner and examine how the learner responds to a variety of questions.

In contrast, the assessment of understanding from the perspective of Dubinsky's APOS theory is less radical constructivist in nature. Rather than relying entirely on interviews, the application of APOS theory according to Asiala et al. (1996) typically utilizes two kinds of tools: written instruments and in-depth interviews. These written instruments include fairly standard items, typically open-ended in nature, focused on the content and the student responses to them receive relatively traditional analysis. These written instruments provide a broad scoped look at what students may or may not be learning and indications about possible mental constructions. From such a broad position, the interviews then narrow in on the displayed understandings by gathering additional information. In particular, Asiala et al. (1996) point out that "the interviews are far more valuable than written assessment instruments used alone ... [since] for one student the written work may

[10]A form-oriented action is a learner's external demonstration helpful in indicating the layer of understanding at which a learner is potentially operating. See Section 2.2 for an enhanced discussion.

appear essentially correct while the [interview] transcript reveals little understanding, while for another student the reverse may be true" (p. 25). As a result, the interviews become tools that permit one to observe the external signs of internal understandings; however, all the data collected in reference to a learner is aggregated to gain a better perspective on the held understandings.

The design and use of the interview questions are another place where the practices of APOS theory significantly diverge from the interview practices associated with Pirie and Kieren's model of understanding. According to Asiala et al. (1996), the interview questions emerge from the genetic decomposition previously developed by the interviewer. Following the interview stage, the analysis of the data feeds into a loop of categorizing student responses, referencing them with respect to the proposed mental constructions defined in the genetic decomposition, and examining the genetic decomposition in light of the data. In particular, Brown et al. (1997) state "student responses are compared to find very fine mathematical points which some students seem to understand (or operations that some can perform) but others cannot. Then we try to find some explanation for the difference in terms of some construction of actions, processes, objects and/or schemas. If we can find an explanation that seems to work, then it is used to revise the genetic decomposition" (p. 189). Thus, the interviews serve a dual purpose of providing insight into the students' mathematical constructions, the mathematics being learned and used, and the descriptive and explanatory powers of the genetic decomposition.

4.5.2. *Instructional practices.* Pirie and Kieren (1994b) note that insights garnered from their model of understanding are useful in the planning and engaging in mathematical lessons as well as making observations about curriculum development. Drawing from their beliefs about students' development of mathematical understanding, they construct a variety of instructional practices that relate the classroom environment to the promotion of growth of understanding. Underlying these practices are a set of beliefs which drive the implementation and interactions of the teacher with the students.

In particular, Pirie and Kieren (1992a) identify four critical tenets teachers must hold when creating a constructivist environment for encouraging mathematical learning and understanding:

> 1) Although a teacher may have the intention to move students toward particular mathematics learning goals, she will be well aware that such progress may not be achieved by some of the students and may not be achieved as expected by others . . .
> 2) In creating an environment or providing opportunities for children to modify their mathematical understanding, the teacher will act upon the belief that there are different pathways to similar mathematical understanding . . .
> 3) The teacher will be aware that different people will hold different mathematical understandings . . .
> 4) The teacher will know that for any topic there are different levels of understanding, but that these are never achieved "once and for all." (pp. 507–508)

The first tenet suggests that a constructivist teacher must be both aware and reactive to the ever-evolving classroom environment. This continual restructuring of the classroom is informed by observations of individual student fabrications of understandings as well as the whole classes' constructions. As a result, a teacher must, according to Kieren (1997), place new emphases:

> (1) on listening to rather than simply listening for;
> (2) on acting with the students in doing mathematics rather than simply showing students how to do mathematics;
> (3) on establishing effective discourse of mathematical argument or mathematical conversation rather than simply the discourse of telling, interrogating, and evaluating;
> (4) on the mechanism of students' mathematical thinking rather than simply on students' answers;
> (5) on the teacher and students as fully implicated by their actions each in learning of the other; and
> (6) on the • teacher [sic] as co-developer of a lived mathematics curriculum not just a recipient of or conduit for a predecided curriculum. (p. 33)

These emphases transform the classroom environment and allow the teacher to focus on the understandings constructed by the individual students while gaining awareness necessary to plan additional classroom experiences.

The second tenet implies that a teacher must recognize there is no specific form or sequence of instruction since no unique or best path to the construction of understanding exists. However, Pirie and Martin (1997) identify a particular effective sequence for teaching the concept of linear equations. This sequence, although presented as applicable across a classroom, involved the teacher with the understandings held by the individual students in the class. Initially, the teacher provided advance organizers to bring the elements of primitive knowing to the forefront of the students' minds and prepare the groundwork for the development of understanding. After this, the teacher presented a task requiring investigation leading students into explorations that would expose the students' thinking, cause reflection, and guide future thinking. The teacher then made constant entreaties for the students to fold back and reorganize lower levels of understanding without inappropriately pushing students. Even though described as a sequence, each student receives different questions and probes; however, this overall instructional practice appears successful (Pirie & Martin, 1997).

The third tenet proposes that the teacher must consider that each individual's understandings are mediated by the internal understandings held by the individual. As a result, the teacher cannot assume that understanding of a certain topic can be transmitted to or gained by students. The reason for this is, according to Pirie and Kieren (1992a), "An understanding of a topic is not an acquisition. Understanding is an ongoing process which is by nature unique to that student" (p. 508). Thus, Pirie and Kieren appear to be stating that the teacher cannot desire to impart a particular idealized version of understanding to students.

The last tenet involves the responses of the teacher regarding observations drawn in the classroom. Even though outward evidences appear to indicate multiple students hold similar understandings, each student has unique understandings. As a result, the teacher must validate each student's level of understanding and compare the levels of understanding across students. Additionally, instruction should not focus solely on the ways of thinking correspondent to a single layer since the normal growth process requires students to systematically fold back to earlier layers of understanding. As a result, the learning environment should be constructed to promote folding back in hopes of advancing students' growth of understanding.

Underlying this entire discussion has been a belief that "teachers cannot give students understanding. Only the student can build her own understanding. The role of the teacher is to provoke and enable this growth" (Pirie & Kieren, 1991a, p. 5). The provocation and enabling of growth reaches beyond simply asking students to work on high level mathematics and includes the generation of opportunities to promote understanding. The three types of questions, provocative, invocative, and validating, are an integral part of this undertaking. By using these questioning techniques, the teacher can address the individualized understandings of a student by provoking movement to an outer layer of understanding, by invoking folding back to a previous level of understanding, and by encouraging students to validate their own reasoning (Pirie & Kieren, 1990).

The instructional approach for fostering conceptual thinking connected with Dubinky's APOS theory differs from the organization provided by Pirie and Kieren although it contains many of the same features. According to Cottrill et al. (1996), "The main contribution it [the APOS theoretical perspective] makes to the design of instruction is to suggest specific mental constructions that can be made in learning the material. The instruction focuses on getting students to make these constructions" (p. 169). As a result, the genetic decomposition represents a possible path a learner may take to the development of conceptual understanding as well as a guide for the development of any instructional activity. Specifically, the genetic decomposition provides the teacher with a general path which may lead the student to construct appropriate understandings. However, Dubinsky et al. (1994) do not consider the development of an instructional sequence as a simplistic, linear sequence where a complex topic is dissected into a logically coherent sequence of small components. In effect, Asiala et al. (1996) point out that according to APOS theory "the growth of understanding is highly non-linear with starts and stops; the student develops partial understandings, repeatedly returns to the same piece of knowledge, and periodically summarizes and ties ideas together" (p. 13). Thus, pedagogical practices need to account for a learner's simultaneously construction of understanding of multiple concepts with one concept subordinate to another. In order to accomplish this, instructional experiences from the APOS perspective employ the holistic spray.

> The idea [of holistic spray] is that everything is sprayed at them in a holistic manner. Each individual (or team) tries to make sense out of the situation—that is, they try to do the problems that the teacher asks them to solve, or to answer questions which the teacher or fellow students ask. In this way the students enhance their understandings of one or another concept bit by bit. They keep coming at it, always trying to make more sense, always learning a little more, and sometimes feeling a great deal of frustration. And it is the role of the teacher, not to eliminate this frustration, but to help students learn to manage it, and to use it as a hammer to smash their own ignorance. (Dubinsky et al., 1994, p. 300)

An integral part of this strategy involves the engagement of students in intentionally disequilibrating environments integrating as much as possible about the concept under study. Of particular importance is the social aspect of the learning situation (Asiala et al., 1996). Rather than the environment supporting individualized exploration, the individuals comprising a cooperative learning group each bring their own perspective to the material presented and therefore construct their own versions of understanding. However, in order for the group to efficiently communicate, members must mediate their understandings to the group during discussion. Therefore, according to Dubinsky's perspectives from APOS theory, the social context in which the learning takes place enhances the individualized construction of understanding.

One particular instantiation of the above theoretical pedagogical approach has been coined by APOS researchers as the *ACE Teaching Cycle* which integrates activities, class discussions, and exercises. The design of the pedagogical approach focuses on eliciting the specific mental constructions suggested by the theoretical analysis. In order to accomplish this, the activities involve cooperative learning groups engaged in the exploration of mathematical concepts using computer assignments as a means of building a knowledge base. In particular, Asiala et al. (1996) stated "It is important to note that there are major differences between these computer activities and the kind of activities used in 'discovery learning.' While some computer activities may involve an element of discovery, their primary goal is to provide students with an experience base rather than to lead them to correct answers" (p. 14). In effect, the computer activities serve as a means of helping students make sense of different portions of the whole concept thereby incrementally setting the stage for enhancing the student's understanding through personal reflection and group discussion. In addition, the computer activities bring to the foreground the components needed to make the theoretical analysis' requisite mental constructions.

To further build upon this foundation laid by the computer activities, the classroom discussions reflect on the computer activities and involve students, working in teams, with paper-and-pencil tasks (Asiala et al., 1996). The instructor actively engages groups in discussion of the computer experiences and how the paper-and-pencil work integrates with those experiences and occasionally providing definitions, explanations, and integrating overviews to move the discussion forward. Finally, in the *ACE Teaching Cycle*, the exercises assigned after the conclusion of the computer activities and classroom discussion are considered relatively traditional but with the intention to "reinforce the ideas they [the students] have constructed, to use the mathematics they have learned and on occasion, to begin thinking about situations that will be studied later" (Asiala et al., 1996, p. 14). One major difference from the exercises contained in a traditional text and the texts born out of the APOS theory is that in traditional texts the exercises provide template problems which correspond to the examples and theory presented in a particular chapter. This template, from the perspective of APOS theory, circumvents the disequilibrium and formation of rich mental constructions considered necessary for the building of understanding. As a result, APOS theory-based texts incorporate the exercises after the computer activities and discussion of the mathematics contained therein. The texts do not provide "worked examples" forcing students to investigate the ideas underlying an activity, reflect upon the computer activities, and integrate those with class discussions to look for linkages and solution paths.

5. Conclusion

Although there exist many theoretical frameworks of mathematical understanding, it appears that both Pirie and Kieren's model of the growth of mathematical understanding and Dubinsky's APOS theory include the salient features of many of them. Pirie and Kieren (1989) posit understanding as a connective, multi-layered, non-linear, recursive process with fractal characteristics. From this perspective, the development of understanding involves the building and re-organization of one's knowledge structures. Movement between layers of understanding is a result of generalization beyond the previous level and the reconstruction of a lower layer understanding. Such movement is evoked by folding back. The building of understanding entails the examination of connections built between concept images, the location of missing, incorrect or partial connections, and the re-organization of these connections into a consistent stable structure. Similarly, Dubinsky asserts understanding to be entwined in schema development thereby requiring the various constructions specified by reflective abstraction. From this perspective, the development of understanding is a continual, although non-linear, process of constructing schemas of greater elaboration which coalesce actions into processes and these processes become encapsulated into objects upon which new actions can take place. The schema organizes these actions, process, objects and subordinate schemas into a structure that transcends the components while providing meaning to a concept.

This paper sought to address the relationships of these two theories to both historical and recent frameworks of mathematical understanding, to identify the elements and constructs of the theories, to examine their roots, to hypothesize how their organizations are related, and to discuss the implications these theories have on assessment and instructional practices. Neither theory has been championed over the other since each theory has its own set of perspectives as well as its own set of strengths. Selecting any theory over another requires one to examine a theory's perspectives. This paper sought to provide a coherent discussion of two theories by comparing and contrasting them at many different levels and from different lenses. As a result, this paper answered a variety of questions; but in doing so, gave rise to several more questions.

First from a philosophical perspective, how can Pirie and Kieren's model of the growth of understanding and Dubinsky's APOS theory simultaneously inform research? Second, if one examined the same interview session from the perspectives of Pirie and Kieren's model of understanding and Dubinsky's APOS theory, what aspects would be commonly uncovered? Uncovered by one but not the other? What instructional implications would be drawn? Third, what operational differences would be evident in a classroom based upon Pirie and Kieren's model of understanding and a classroom based upon Dubinsky's APOS theory? Would the goals of the teacher change? Would the differences significantly affect instructional practice, assessment procedures and interpretation, attitudes toward standardized testing, or interaction with the students? This paper cannot hope to answer these questions; however, perhaps with future research answers will become available.

References

Asiala, M., Brown, A., DeVries, D. J., Dubinsky, E., Mathews, D., & Thomas, K. (1996). A framework for research and curriculum development in Undergraduate mathematics education. In J. Kaput, A. H. Schoenfeld & E. Dubinsky (Eds.) *Research in Collegiate Mathematics Education* (pp. 1–32). Providence, RI: American Mathematical Society.

Asiala, M., Brown, A., Kleiman, J., & Mathews, D. (1998). The development of students' understanding of permutations and symmetries. *International Journal of Computers for Mathematical Learning, 3*, 13–43.

Ayers, T., Davis, G., Dubinsky, E., & Lewin, P. (1988). Computer experiences in learning composition of functions. *Journal for Research in Mathematics Education, 19(3)*, 246–259.

Bachelard, G. (1938). *La formation de l'esprit scientifique.* Paris: Vrin.

Ball, D. (1991). Research on teaching mathematics: Making subject matter knowledge part of the equation. In J. Brophy (Ed.), *Advances in research on teaching: Vol. 2 Teachers' subject matter knowledge and classroom instruction* (pp. 1–48). Greenwich, CT: JAI Press.

Brown, A., DeVries, D. J., Dubinsky, E. & Thomas, K. (1997). Learning binary operations, groups, and subgroups. *Journal of Mathematical Behavior, 16*(3), 187–239.

Brown, J. S., Collins, A., & Duguid, P. (1989). Situated cognition and the culture of learning. *Educational Researcher, 18*(1), 32–42.

Brownell, W. A. (1945) Psychological considerations in the learning and teaching of arithmetic. In W. D. Reeve (Ed.), *The teaching of arithmetic. Tenth yearbook of the National Council of Teachers of Mathematics* (pp. 1–31). New York: Teachers College, Columbia University.

Brownell, W. A. & Sims, V. M. (1946). The nature of understanding. In J. F. Weaver & J. Kilpatrick (Eds.) (1972), *The place of meaning in mathematics instruction: Selected theoretical papers of William A. Brownell* (Studies in Mathematics, Vol. 21, pp. 161–179). Stanford University: School Mathematics Study Group. (Originally published in *The measurement of understanding, Forty-fifth Yearbook of the National Society for the Study of Education*, Part I, 27–43.)

Burton, L. (1984). Mathematical thinking: The struggle for meaning. *Journal for Research in Mathematics Education, 15*(1), 35–49.

Byers, V. & Erlwanger, S. (1985). Memory in mathematical understanding. *Educational Studies in Mathematics, 16*, 259–281.

Byers, V. & Herscovics, N. (1977). Understanding school mathematics. *Mathematics Teaching, 81*, 24–27.

Clark, J., Cordero, F., Cottrill, J., Czarnocha, B., DeVries, D., St. John, D., Tolias, G. & Vidakovic, D. (1997). Constructing a schema: The case of the chain rule. *Journal for Mathematical Behavior, 16*(4), 345–364.

Cornu, B. (1991). Limits. In D. Tall (Ed.), *Advanced Mathematical Thinking* (pp. 153–166). Dordrecht: Kluwer.

Cottrill, J. (1999). *Students' understanding of the concept of chain rule in first year calculus and the relation to their understanding of composition of functions.* Unpublished doctoral dissertation, Purdue University.

Cottrill, J., Dubinsky, E., Nichols, D., Schwingendorf, K., Thomas, K. & Vidakovic, D. (1996). Understanding the limit concept: Beginning with a coordinated process scheme. *Journal of Mathematical Behavior, 15*, 167–192.

Davis, E. J. (1978, September). A model for understanding in mathematics. *Arithmetic Teacher*, 13–17.

Davis, R. B. (1984). *Learning mathematics: The cognitive science approach to mathematics education.* Norwood, NJ: Ablex.

Davis, R. B. & Vinner, S. (1986). The notion of limit: Some seemingly unavoidable misconception stages. *Journal of Mathematical Behavior, 5*, 281–303.

Driver, R. & Easley, J. (1978). Pupils and paradigms. *Studies in Science Education, 5*, 61–84.

Dubinsky, E. (1991). Reflective abstraction in advanced mathematical thinking. In D. Tall (Ed.), *Advanced mathematical thinking* (pp. 95–123). Dordrecht: Kluwer.

Dubinsky, E. (1994). A theory and practice of learning college mathematics. In A. Schoenfeld (Ed.), *Mathematical thinking and problem solving* (pp. 221–247). Hillsdale, NJ: Erlbaum.

Dubinsky, E., Dautermann, J., Leron, U., & Zazkis, R. (1994). On learning fundamental concepts of group theory. *Educational Studies in Mathematics, 27*, 267–305.

Dubinsky, E. & McDonald, M. (2001). *APOS: A constructivist theory of learning in undergraduate mathematics education research.* In D. Holton et al. (Eds.), *The teaching and learning of mathematics at university level: An ICMI study* (pp. 273–280). Netherlands: Kluwer Academic Publishers.

Fehr, H. (1955). A philosophy of arithmetic instruction. *Arithmetic Teacher, 2*, 27–32.

Ferrini-Mundy, J. & Gaudard, M. (1992). Secondary school calculus: Preparation or pitfall in the study of college calculus? *Journal for Research in Mathematics Education, 23*(1), 56–71.

Flavell, J. H. (1977). *Cognitive development.* Englewood Cliffs, New Jersey: Prentice Hall.

Ginsburg, H. P., Lopez, L.S., Mukhopadhyay, S., Yamamota, T., Willis, M. & Kelley, M. S. (1992). Assessing understandings of arithmetic. In R. Lesh & S. Lamon (Eds.), *Assessment of authentic performance in school mathematics* (pp. 265–289). Washington, DC: AAAS Press.

Greeno, J. G. (1977). Process of understanding in problem solving. In N.J. Castellan, Jr., D. B. Pisoni, & G. R. Potts (Eds.), *Cognitive theory* (Vol. 2, pp. 43–83). Hillsdale, NJ: Erlbaum.

Greeno, J. G. (1991). Number sense as situated knowing in a conceptual domain. *Journal for Research in Mathematics Education, 22*(3), 170–218.

Grinevitch, O. (in preparation). *Student understanding of abstract algebra: A theoretical examination.* Unpublished doctoral dissertation, Bowling Green State University.

Harel, G. & Tall, D. (1991). The general, the abstract, and the generic in advanced mathematics. *For the Learning of Mathematics, 11*(1), 38–42.

Hatano, G. (1988). Social and motivational bases for mathematical understanding. In G. B. Saxe & M. Gearhart (Eds.), *New Directions for Child Development* (Vol. 41, pp. 55–70).

Herscovics, N. & Bergeron, J. (1988). An extended model of understanding. *Proceedings of PME-NA 10* (pp. 15–22). Dekalb, IL: Northern Illinois University, .

Hiebert, J. (Ed.) (1986). *Conceptual and procedural knowledge: The case of mathematics.* Hillsdale, NJ: Erlbaum.

Hiebert, J. & Carpenter, T. P. (1992). Learning and teaching with understanding. In D. Grouws (Ed.), *Handbook of research on mathematics teaching and learning* (pp. 65–97). New York, NY: Macmillan.

Hiebert, J. & Lefevre, P. (1986). Conceptual and procedural knowledge in mathematics: An introductory analysis. In J. Hiebert (Ed.), *Conceptual and procedural knowledge: The case of mathematics* (pp. 1–27). Hillsdale, NJ: Erlbaum.

Hiebert, J. & Wearne, D. (1986). Procedures over concepts: The acquisition of decimal number knowledge. In J. Hiebert (Ed.), *Conceptual and procedural knowledge: The case of mathematics* (pp. 199–223). Hillsdale, NJ: Erlbaum.

Janvier, C. (Ed.) (1987). *Problems of representation in the teaching and learning of mathematics.* Hillsdale, NJ: Erlbaum.

Kaput, J. (1985). Representation and problem solving: Methodological issues related to modeling. In E. A. Silver (Ed.), *Teaching and learning mathematical problem solving: Multiple research perspectives* (pp. 381–398). Hillsdale, NJ: Erlbaum.

Kaput, J. (1987a). Representation and mathematics. In C. Janvier (Ed.), *Problems of representation in mathematics learning and problem solving* (pp. 19–26). Hillsdale, NJ: Erlbaum.

Kaput, J. (1987b). Towards a theory of symbol use in mathematics. In C. Janvier (Ed.), *Problems of representation in mathematics learning and problem solving* (pp. 159–195). Hillsdale, NJ: Erlbaum.

Kaput, J. (1989a). Linking representations in the symbol systems of algebra. In C. Kieran & S. Wagner (Eds.), *A research agenda for the learning and teaching of algebra* (pp. 167–194). Reston, VA: National Council of Teachers of Mathematics; Hillsdale, NJ: Erlbaum.

Kaput, J. (1989b). Supporting concrete visual thinking in multiplicative reasoning: Difficulties and opportunities. *Focus on Learning Problems in Mathematics, 11*(1), 35–47.

Kaput, J. (1992). Technology and mathematics education. In D. Grouws (Ed.), *Handbook of research on mathematics teaching and learning* (pp. 515–556). New York, NY: Macmillan

Kieren, T. E. (1990). Understanding for teaching for understanding. *The Alberta Journal of Educational Research, 36*(3), 191–201.

Kieren, T. E. (1997). Theories for the classroom: Connections between research and practice. *For the Learning of Mathematics, 17*(2), 31–33.

Lehman, H. (1977). On understanding mathematics. *Educational Theory, 27*, 2.

Leinhardt, G. (1988). Getting to know: Tracing student's mathematical knowledge from intuition to competence. *Educational Psychologist, 23*(2), 119–144.

Mack, N. (1990). Learning fractions with understanding: Building on informal knowledge. *Journal for Research in Mathematics Education, 21*(1), 16–32.

Mansfield, H. M. & Happs, J. C. (1992). Using grade eight students' existing knowledge to teach about parallel lines. *School Science and Mathematics, 92*(8), 450–454.

McDonald, M. A., Mathews, D. M. & Strobel, K. H. (2000). Understanding sequences: A tale of two objects. In E. Dubinsky, A. H. Schoenfeld & J. Kaput (Eds.), *Research in collegiate mathematics education* (pp. 77–102). Providence, RI: American Mathematical Society.

McLellan, J. A. & Dewey, J. (1895). *The psychology of number and its applications to methods of teaching arithmetic.* New York: D. Appleton.

Michener, E. R. (1978). Understanding understanding mathematics. *Cognitive Science, 2*, 361–383.

Nesher, P. (1986). Are mathematical understanding and algorithmic performance related? *For the Learning of Mathematics, 6*(3), 2–9.

Nickerson, R. S. (1985). Understanding understanding. *American Journal of Education, 93*(2), 201–239.

Ohlsson, S. (1988). Mathematical meaning and applicational meaning in the semantics of fractions and related concepts, In J. Hiebert & M. Behr (Eds.), *Number concepts and operations in the middle grades* (pp. 53–91). Reston, VA: National Council of Teachers of Mathematics.

Ohlsson, S., Ernest, A. M., & Rees, E. (1992). The cognitive complexity of learning and doing arithmetic. *Journal for Research in Mathematics Education, 23*(5), 441–467.

Piaget, J. (1970). *Genetic epistemology.* New York, NY: W.W. Norton.

Piaget, J. (1975). Piaget's theory. In P. B. Neubauer (Ed.), *The process of child development* (pp. 164–212). New York, NY: John Aronson.

Pirie, S. E. B. (1988). Understanding: Instrumental, relational, intuitive, constructed, formalized ... ? How can we know? *For the Learning of Mathematics, 8*(3), 2–6.

Pirie, S. E. B. & Kieren, T. E. (1989). A recursive theory of mathematical understanding. *For the Learning of Mathematics, 9*(3), 7–11.

Pirie, S. E. B. & Kieren, T. E. (1990). *A recursive theory for the mathematical understanding—some elements and implications.* Paper presented at the Annual Meeting of the American Educational Research Association (Boston, MA, April 1990).

Pirie, S. E. B. & Kieren, T. E. (1991a). *A dynamic theory of mathematical understanding: Some features and implications.* (ERIC Document Reproduction Service No. ED 347 067)

Pirie, S. E. B. & Kieren, T. E. (1991b). *Folding back: Dynamics in the growth of mathematical understanding.* Paper presented at the Fifteenth Meeting of the Psychology of Mathematics Education Conference (Assissi, Italy, July 1991).

Pirie, S. E. B. & Kieren, T. E. (1992a). Creating constructivist environments and constructing creative mathematics. *Educational Studies in Mathematics, 23*, 505–528.

Pirie, S. E. B. & Kieren, T. E. (1992b). Watching Sandy's understanding grow. *Journal of Mathematical Behavior, 11*, 243–257.

Pirie, S. E. B. & Kieren, T. E. (1994a). Beyond metaphor: Formalizing in mathematical understanding with constructivist environments. *For the Learning of Mathematics, 14*(1), 39–43.

Pirie, S. E. B. & Kieren, T. E. (1994b). Growth in mathematical understanding: How can we characterise it and how can we represent it? *Educational Studies in Mathematics, 26*, 165–190.

Pirie, S. E. B. & Martin, L. (1997). The equation, the whole equation and nothing but the equation! One approach to the teaching of linear equations. *Educational Studies in Mathematics, 34*, 159–181.

Pirie, S. E. B. & Schwarzenberger, R. L. E. (1988). Mathematical discussion and mathematical understanding. *Educational Studies in Mathematics, 19*, 459–470.

Polya, G. (1945) *How to solve it.* Princeton, NJ: Princeton University Press.

Polya, G. (1962). *Mathematical discovery* (Vol. 2). New York, NY: Wiley.

Resnick, L. B. & Omanson, S. (1987). Learning to understand arithmetic. In R. Glaser (Ed.), *Advances in instructional psychology* (Vol. 3, pp. 41–95). Hillsdale, NJ: Erlbaum.

Resnick, L. B., Nesher, P., Leonard, F., Magone, M., Omanson, S., & Peled, I. (1989). Conceptual bases of arithmetic errors: The case of decimal fractions. *Journal for Research in Mathematics Education, 20*(1), 8–27.

Saxe, G. B. (1988). Studying working intelligence. In B. Rogoff & J. Lave (Eds.), *Everyday cognition* (pp. 9–40). Cambridge, MA: Harvard University Press.

Schoenfeld, A. H. (1989). Exploring the process problem space: Notes on the description and analysis of mathematical processes. In C. Maher, G. Goldin, & R. B. Davis (Eds.), *Proceedings of psychology of mathematics education North America XI* (Vol. 2, pp. 95–120). New Brunswick, NJ: Rutgers, Centre for Mathematics, Science and Computer Education.

Schoenfeld, A. H. (1998). Toward a theory of teaching-in-context. *Issues in Education, 4*(1), 1–94.

Schoenfeld, A. H. (2000). Purposes and methods of research in mathematics education. *Notices of the American Mathematical Society, 47*(6), 641–649.

Schroeder, T. L. (1987). Student's understanding of mathematics: A review and synthesis of some recent research. In J. Bergeron, N. Herscovics, & C. Kieran (Eds.), *Psychology of Mathematics Education XI* (Vol. 3, pp. 332–338). Montreal: PME.

Sfard, A. (1991). On the dual nature of mathematical conceptions: Reflections on processes and objects as different sides of the same coin. *Educational Studies in Mathematics, 22*, 1–36.

Sfard, A. (1992). Operational origins of mathematical objects and the quandry of reification—The case of function. In G. Harel & E. Dubinsky (Eds.), *The concept of function: Aspects of epistomology and pedagogy* (pp. 59–84). Washington, DC: MAA.

Sfard, A. (1994). Reification as the birth of metaphor. *For the Learning of Mathematics, 14*(1), 44–55.

Sfard, A. & Linchevski, L. (1994). The gains and the pitfalls of reification—The case of algebra. *Educational Studies in Mathematics, 26*, 191–228.

Skemp, R. R. (1976). Relational understanding and instrumental understanding. *Mathematics Teaching, 77*, 20–26.

Skemp, R. R. (1979). Goals of learning and qualities of understanding. *Mathematics Teaching, 88*, 44–49.

Skemp, R. R. (1982). Symbolic understanding. *Mathematics Teaching, 99*, 59–61.

Sierpinska, A. (1987). Humanities students and epistemological obstacles related to limits. *Educational Studies in Mathematics, 18*, 371–397.

Sierpinska, A. (1990a). *Remarks on understanding in mathematics.* Paper presented at the 1990 meeting of the Canadian Mathematics Education Study Group (Vancouver, CAN, 1990).

Sierpinska, A. (1990b). Some remarks on understanding in mathematics. *For the Learning of Mathematics*, *10*(3), 24–41.

Silver, E. A. (1986). Using conceptual and procedural knowledge: A focus on relationships. In J. Hiebert (Ed.), *Conceptual and procedural knowledge: The case of mathematics* (pp. 181–189). Hillsdale, NJ: Erlbaum.

Simon, M. A. (1993). Prospective elementary teachers' knowledge of division. *Journal for Research in Mathematics Education*, *24*(3), 233–254.

Steffe, L. P. & Kieren, T. E. (1994). Radical constructivism and mathematics education. *Journal for Research in Mathematics Education*, *25*(6), 711–733.

Tall, D. (1978). The dynamics of understanding mathematics. *Mathematics Teaching*, *84*, 50–52.

Tall, D. (1989). Concept images, generic organizers, computers, and curriculum change. *For the Learning of Mathematics*, 9(3), 37–42.

Tall, D. (1991). The psychology of advanced mathematical thinking. In D. Tall (Ed.), *Advanced mathematical thinking* (pp. 3–21). Dordrecht: Kluwer.

Tall, D. & Vinner, S. (1981). Concept image and concept definition in mathematics with special reference to limits and continuity. *Educational Studies in Mathematics*, *12*, 151–169.

Towers, J., Martin, L. & Pirie, S. E. B. (2000). Growing mathematical understanding: Layered observations. In M. L. Fernandez (Ed.), *Proceedings of the Twenty-Second Annual Meeting of the North American Chapter of the International Group for the Psychology of Mathematics Education* (pp. 225–230). Columbus, OH: ERIC Clearinghouse.

Van Engen, H. (1949). An analysis of meaning in arithmetic. *Elementary School Journal*, *49*, 321–329; 395–400.

VanLehn, K. (1986). Arithmetic procedures are induced from examples. In J. Hiebert (Ed.), *Conceptual and procedural knowledge: The case of mathematics* (pp. 133–180). Hillsdale, NJ: Erlbaum.

Vinner, S. (1983). Concept definition, concept image and the notion of function. *International Journal of Mathematical Education in Science and Technology*, *14*, 293–305.

Vinner, S. (1991). The role of definitions in the teaching and learning of mathematics. In D. Tall (Ed.), *Advanced mathematical thinking* (pp. 65–81). Dordrecht: Kluwer.

von Glasersfeld, E. (1987). Learning as a constructive activity. In C. Janvier (Ed.), *Problems of representation in the learning and teaching of mathematics* (pp. 3–18). Hillsdale, NJ: Erlbaum.

Wearne, D. & Hiebert, J. (1988). A cognitive approach to meaningful mathematics instruction: Testing a local theory using decimal numbers. *Journal for Research in Mathematics Education*, *19*(5), 371–384.

Wertheimer, M. (1959). *Productive thinking*. New York: Harper & Row.

Department of Mathematics and Statistics, Bowling Green State University, Bowling Green OH 43403

E-mail address: meel@bgnet.bgsu.edu

CBMS Issues in Mathematics Education
Volume **12**, 2003

The Nature of Learning in Interactive Technological Environments: A Proposal for a Research Agenda Based on Grounded Theory

Jack Bookman and David Malone

Abstract. The purpose of this study is to develop, based on observations of students' work, a set of research questions that will help us understand how students learn in a particular technology-rich environment, one using a computer algebra system together with an HTML document using text, hyperlinks, and Java applets. These questions were not derived a priori from a theoretical perspective but were derived from the data. From the data we identified three categories of research questions: (1) What is the role of the instructor in this environment? (2) What types of behavior and thinking processes are students engaged in as they work together in front of the computer, and how can the modules be written to facilitate students' self-monitoring and effective collaborative interaction? and (3) What opportunities and obstacles are raised by the technology itself? Research in each of these areas has important implications for curriculum developers, mathematics instructors, and students.

Introduction

It is clear that technology is fundamentally changing the way we live, work, and learn, but is not clear exactly how it is changing the way we live, work, and learn. In particular, it is not clear how the Internet and sophisticated computer algebra systems are changing and will change how we teach and learn mathematics. Smith (2000) has succinctly summed up the situation:

> Technology is a fact of life for our students—before, during, and after college. Most students entering college now have experience with a graphing calculator, because calculators are permitted or required on major standardized tests. A large and growing percentage of students have computer experience as well—at home, in the classroom, or in a school or public library. "Surfin' the 'Net" is a way of life—whether for good reasons or bad. Many

This research was funded by National Science Foundation grant # DUE-9752421. The authors would also like to acknowledge the support for this project from the Carnegie Academy for the Scholarship of Teaching and Learning, a program of The Carnegie Foundation for the Advancement of Teaching.

> colleges require computer purchase or incorporate it into their tuition structure. Where the computer itself is not required, the student is likely to find use of technology required in a variety of courses. After graduation, it is virtually certain that, whatever the job is, there will be a computer close at hand. And there is no sign that increase in power or decrease in cost will slow down any time in the near future. We know these tools can be used stupidly or intelligently, and intelligent choices often involve knowledge of mathematics, so this technological environment is our business. Since most of our traditional curriculum was assembled in a pre-computer age, we have a responsibility to rethink whether this curriculum still addresses the right issue in the right ways—and that is exactly what has motivated some reformers.

Nonetheless, the move to integrate technology into teaching has not been without its detractors. Krantz (2000) some important concerns when he states that: (1) distance education and products promoted by publishers for profit "describe a dangerous trend"; (2) "Provosts and deans have dollars signs in their eyes. They envision teaching more students with fewer faculty;" and (3) "The important question is whether students are internalizing and retaining the material." But he presents an extreme either/or view of technology lumping together all use of technology in the classroom. He asserts, providing no evidence, that "Traditional education . . . enables students to master the ideas and retain them for future use" and that "traditional methods . . . have had—and continue to have—great success," that traditional classrooms produce "interaction of first rate minds." He then claims (again with no evidence) that there is no measurable benefit to employing technology in the classroom. He is not the only mathematician with these concerns. The issue of how, or if, to introduce technology into the classroom is one of the most divisive and emotionally charged issues in education.

In this paper, we propose an agenda for research that will move the discussion of the use of technology in mathematics classes from the coffee lounge and soap box to the seminar room. We will discuss some preliminary results of careful observations of student learning using computer algebra systems with lessons delivered via the Internet. Based on these observations, we propose a set of research questions whose answers will help us to understand how best to use these new technologies to improve the teaching and learning of mathematics.

Background

In recent years, consensus among organizations concerned about mathematics education, such as the National Council of Teachers of Mathematics and National Research Council, may have emerged regarding the essential steps in reforming mathematics and science education (Battista, 1999; Chambers & Bailey, 1996; National Council of Teachers of Mathematics, 1991; National Research Council, 1991; National Science Foundation, 1996). Bailey and Chambers (1996) summarized several of these reform reports and concluded that six overarching recommendations have emerged: (1) integrate the teaching of science and mathematics; (2) emphasize cooperative learning; (3) focus on application and relevant problem solving;

(4) teach primarily through active learning as opposed to lecture; (5) attend to the motivation of learners; and (6) use technology in meaningful ways.

The Connected Curriculum Project[1] is an innovative instructional effort which addresses each of these six recommendations. The Connected Curriculum Project (CCP) has developed a collection of learning materials designed to create interactive learning environments for students in the first two years of college mathematics courses (Colvin et al., 1999; Coyle et al., 1998). The materials combine the interactivity, accessibility, and connectivity of the Web with the power of computer algebra systems. These materials may be used by groups of students as an integrated part of a course, by individuals as independent projects, or as supplements to classroom discussions. Lawrence Moore and David Smith, who began their collaboration in 1988 at the beginning of the calculus reform movement, lead this project.

The CCP is a direct extension of the experience gained from the calculus reform movement in general and, in particular, Project CALC: Calculus as a Laboratory Course, supported by the NSF Calculus Reform Initiative. The key features of that course are real-world problems, hands-on activities, discovery learning, writing and revision of writing, teamwork, and intelligent use of available tools. The stated goals for the course are that students should: (1) be able to use mathematics to structure their understanding of and investigate questions in the world around them; (2) be able to use calculus to formulate problems, to solve problems, and to communicate the solution of problems to others; (3) be able to use technology as an integral part of this process of formulation, solution, and communication; and (4) learn to work cooperatively (Bookman & Blake, 1996; Smith & Moore, 1990, 1991).

A subsequent NSF grant in 1993 (Interactive Modules for Courses Following Calculus, Duke University, NSF DUE-9352889, 1993–97) supported development of modular lab activities for courses beyond calculus: linear algebra, differential equations, and engineering mathematics. These modules were created as interactive texts in specific computer algebra systems. CCP was devised to extend the usefulness of these modules by capitalizing on the interactivity and availability provided by the Internet. The CCP modules include hypertext links, Java applets, sophisticated graphics, a computer algebra system, realistic scenarios, and questions that require written answers. The materials used for this study were single-topic units that can be completed in one to two hours with students working in two-person teams in a computer lab environment.

The CCP is based in part on ideas of cognitive psychology that examine the ways in which students take in, organize, and represent knowledge internally. An underlying principle from this research is that students cannot simply be given knowledge; they must construct knowledge in their own minds. This perspective on learning, known as constructivism, is rooted in the earlier work of cognitive theorists such as Piaget, Bruner, and Vygotsky (Bruner, 1996; Piaget, 1952; Vygotsky, 1978).

Learning from the constructivist perspective is seen as a "self-regulated process of resolving inner cognitive conflicts that often become apparent through concrete experience, collaborative discourse, and reflection" (Fosnot, 1993). constructivist theorists maintain that the active learning in a socially interactive environment is a necessary condition for meaningful understanding. The primary role of the

---

[1]The Coordinated Curriculum Library, Duke University, NSF DUE-9752421, 1998-2001. The Connected Curriculum Project was funded by a granted entitled The Coordinated Curriculum Library

teacher is to structure learning situations in which students experience a sense of cognitive disequilibrium. Students take on more responsibility for monitoring and regulating their own thinking and learning. The job of the teacher is to create learning environments that place students in a position of constructing or building meaning and understanding for themselves. Thus, from this viewpoint, learning is viewed as a transforming process involving conceptual change, not merely a process in which students recite back information that they have passively accumulated.

In the traditional mathematics class, the instructor might explain a concept, demonstrate several examples of how to solve problems involving that particular concept, and then ask students individually to work through problem sets. The emphasis is on learning computational procedures. In a cognitively oriented or constructivist mathematics class, after an introductory discussion led by the teacher, students are given complex and engaging problems which are situated and embedded in a meaningful context. These contextualized problems require students to engage not only in computational procedures, but also in sustained mathematical reasoning. Students might be asked to write about the problem, create graphs and drawings, manipulate objects, and in other ways actively make sense of the problem. Working collaboratively with peers, they pose hypotheses, justify ideas, formulate solutions, and explain their personal understandings in their own words. An emphasis in a constructivist mathematics class is on making sense of mathematical ideas.

Because constructivist approaches to instruction attempt to engage students in solving meaningful problems using real world data, the use of technology in mathematics instruction is seen as holding enormous promise. Portela (1999) "Although not new, constructivism has more relevance in education today because the dawn of the Information Age has rapidly increased the amount of, and accessibility to, information." He stated that there is a scarcity of studies about how students learn in technology-based environments and describes the results of his case study of a mathematical communication and technology course for mathematics graduate students. He reported that the focus of teaching and learning shifted from knowledge transmission to knowledge building and he credits the Internet with aiding this shift. He also indicated that being connected to the Internet in the classroom provided opportunities for more active learning by encouraging students to learn by doing, concentrate on the subject matter rather than simply copying notes from the board, participate in class discussions, work at their own pace, receive individual help from the instructor without holding back the rest of the class, and access related sites "right on the spot."

Papert (1980) has written extensively about ways technology can be used in mathematics classrooms to promote "agency" or student ownership of mathematical thinking. Papert indicated that technology-rich environments which utilize inquiry-based approaches to learning math have the potential of significantly altering the dynamics between instructor and student, as well as between the student and the mathematical content being studied. Papert also noted that in such learning environments students exercise more authority over their own thinking and develop a deeper intuition for mathematical problem solving.

Cooper (1999) stated that, "In its use as an educational medium in a carefully structured learning environment based on the principles of cognitive research, the computer may serve as a strong mechanism for reorganizing mental processes, aiding students in developing the hierarchical structure for their new knowledge." She also noted that although original uses of the computer focused on drill, practice, and individualized learning, use of the computer as an instructional tool is most effective in collaborative learning environments.

The nature of learning mathematics interactively and in technology rich environments has been the focus of several researchers. Dubinsky and Schwingendorf (1997) investigated the effectiveness of teaching calculus using small cooperative learning groups in a computer laboratory setting. They concluded that the "use of computer activities and small group problem solving to implement a theory of learning mathematics has shown itself to be a very promising direction" (p. 241). Dubinsky and Schwingendorf (1997) also noted concerns such as the need for computer software to be "as easy as possible to use" (p. 235) and the need for instructors to develop mechanisms for ensuring that each student is held individually accountable for participating in the collaborative lab environment.

Asiala and Dubinsky (2000) investigated the attitudes and academic performance of students enrolled in college math courses that utilized "innovative pedagogical strategies" (p. 1). These non-traditional teaching approaches included the use of computers and cooperative learning. Asiala and Dubinsky (2000) concluded that these innovative approaches led to improvement in student learning. The researchers also reported a tendency for students who had successfully completed an innovative math course to take more math courses in future semesters than were taken by students who had completed traditionally taught math courses.

Other researchers, including Davidson (1990) and Dubinsky and Fenton (1996) have examined in depth issues surrounding the use of active learning strategies and student collaboration in college mathematics classrooms. Furthermore, Project CLUME has focused on ways collaborative learning strategies can be used effectively to foster deep understanding of mathematical concepts (Rogers et al., 2001). Research in this area has addressed issues of significance to our current study, such as the question of how to get students to reflect on the quality of their social interactions in collaborative learning situations.

We agree with Pea (1987) that "the computer can serve as a fundamental mediational tool for promoting dialogue and collaboration on mathematical problem solving" (p. 125). Research such as that of Pea and Dubinsky (Asiala & Dubinsky, 2000; Dubinsky & Fenton, 1996; Dubinsky & Schwingendorf, 1997) has raised many of the questions we raise here. But much of this previous work predates HTML. Since HTML is a new and potentially powerful educational tool, it is necessary to reexamine previous work in light of this technology. Although many of the questions we raise are old (important and still open) questions about cooperative learning and technology, we also raise some important new questions particularly concerning issues about students use of HTML. For example (as we will document later in this paper) the following questions arose from observing students working in this environment: What cognitive conditions prompt students to use hot links? What are the ways students seek and receive help in the computer learning environment? How can we get students to use help tools built into computer environments so they spend less time floundering and getting frustrated on syntactical problems? Our research also differs from previous work in that, instead of making observations of students working as a group in a classroom, we collected data in such a way to allow for very close examination and documentation of students' behaviors in a way that has not apeared in the research literature.

Many of the questions raised by the current study on socially interactive, technology-rich approaches to learning are relevant to active learning environments in general. A significant research literature exists on active learning and cooperative learning in mathematics as well as on technology in mathematics education. However, significantly less research has been conducted which closely examines the nature of the interaction of active, collaborative learning environments with technology-rich mathematics learning environments. More research is necessary to understand the ways in which socially interactive approaches to teaching interact with the use of computers in mathematics classrooms. It seems important to identify characteristics which are unique to technology-rich classrooms.

Heid et al. (1998) reviewed the empirical research on mathematics learning using computer algebra systems (CAS). Reporting only on those studies that involved systematic data collection and analysis, Heid et al. identified 64 studies from journal articles, conference proceedings and dissertations that addressed CAS. They examined five sets of outcomes—achievement, affect, behavior, strategies and understanding—and concluded that the research justifies incorporating CAS into the established mathematics curriculum. In particular, the researchers noted that, "The majority of studies examined indicate that there is no loss in proficiency in computational skills and these results are obtained in the absence of a CAS on the research instrument. Cumulatively, these studies suggest that use of a CAS in the learning of mathematics . . . can result in higher overall achievement in terms of both procedural and conceptual items." They concluded, "CAS research is now ready to enter a new phase. Researchers must no longer focus their efforts on corroborating the 'no-harm-done' conclusion. They must no longer be satisfied to establish that conceptual understanding is better. They must, like some of the more recent pioneers, investigate the very nature of learning with CAS."

Data Collection and Analysis

Methods. The purpose of this study is to develop, based on observations of students' work, a set of research questions that will help us understand the nature of learning in these more interactive technological environments. Particularly since little research has been done in this area, this phase of the research must be exploratory in nature. We feel the most appropriate research method for this type of research is Glaser and Strauss's notion of *grounded theory*, which they described as "the discovery of theory from data systematically obtained from social research," which they contrasted with "theory generated by logical deduction from *a priori* assumptions" (Glaser & Strauss, 1967). In the first stage of building a grounded theory, researchers examine their data with the purpose of establishing categories and/or constructs unbiased by prior conceptions. The data are studied "to identify significant phenomena, and then determine which phenomena share sufficient similarities that they can be considered instances of the same concept. . . You will need to define the category, give it a label and specify guidelines that you and others can use to determine whether each segment in the database is or is not an instance of the category" (Gall et al., 1996). Our data gathering methods can be described using Romberg's (1992) method of clinical observations where "the details of what one observes shift from predetermined categories to new categories, depending upon initial observations."

Subjects. The subjects studied were college students taking a mathematics course (at a level beyond calculus) in a major research university. The students had been using CCP modules for at least several weeks and were somewhat familiar with *Maple* (the computer algebra system) and the format of the modules. CCP modules were required for their current mathematics coursework and, on average, these students had completed one module per week. For all but one pair of the subjects, the particular module used in the study was a specific requirement for the course in which they were enrolled. The subjects volunteered to be videotaped for the purposes of this study (and were each paid $25). Their participation in the study consisted of working through one of the CCP modules with a partner. The students working together were videotaped and, simultaneously, their computer output was collected on a separate videotape. Each session was 1–2 hours in length and data were collected from a total of 10 pairs of students.

Data Collection. The data were collected in a quiet office where the pair of students could work comfortably. Also in the room (though not always for the entire time) was one of the investigators. On the table was a computer with *Maple*; students were also given paper and pencil. Also in the room was a video camera to record their work and a scan converter connected to a VCR and television to record their computer output. When the students arrived the investigator explained the general purpose of the research and asked the subjects to sign consent forms and forms to be paid. He then helped them find the URL for the module and asked them to begin their work.

Although (as discussed above) other researchers have investigated issues similar to those addressed in this study, the method of the current study is somewhat different. In the current study, not only were subjects videotaped, their computer output was simultaneously recorded and the focus of the video camera remained on the two students sitting at the computer for the entire session (as opposed to videotaping of a larger classroom). These aspects of the current study's methodology provide the researchers with an opportunity to closely examine and document student behavior.

Analysis. A principle of grounded theory is that one generates conceptual categories from evidence (Glaser & Strauss, 1967). The authors of this paper (one mathematics educator and one educational psychologist) observed each of the tapes several times and noted those issues that apeared to facilitate or inhibit learning or that apeared to be important factors in understanding the process of learning taking place. Another principle of grounded theory is that the categories that "emerged from the data are constantly being selectively reformulated by them. The categories, therefore, will fit the data, be understood both to sociologists and to laypeople who are knowledgeable in the area, and make the theory usable for theoretical advance as well as for practical application" (Glaser & Strauss, 1967). The following three categories emerged: (1) the role of the teacher; (2) types of behavior, thinking processes and self-monitoring as students engage in collaborative interaction; and (3) issues raised directly by the technology. We have selected several excerpts from the videotapes that illustrate these concepts. For each of the vignettes, we discuss aspects of the four categories that are reflected in the data. We then discuss researchable questions raised by our analysis.

In order to help the reader place these vignettes in the context in which these students would normally be working, we present the following description of the beginning of a typical class in which these students would be working on these CCP modules:

> On a day when CCP computer based modules are being used, the learning environment looks significantly different from a traditional mathematics class. Typically, after gaining the attention of students and taking care of administrative announcements, the instructor gives a brief overview of and introduction to the lesson. The purpose of this initial teacher-directed overview is to activate students' prior knowledge, introduce new terminology and procedures, and to provide students with a conceptual anchor. Students are then assigned to pairs (or get with a previously assigned partner). For most CCP modules, students work cooperatively with a single partner, but each pair of students belongs to a larger support team made up of four students. Each pair works collaboratively on a single computer. Roles are not assigned, so students must decide between themselves who will control the keyboard, point the mouse, read the problem, and take primary responsibility for the variety of tasks required by the learning activity. Once settled in, each of the two students typically reads the introduction on the computer screen and then they collaboratively engage in problem solving.

Vignettes

Vignette 1. In this vignette we examine the work of Mary and Jim on a module called the "The Equiangular Spiral"[2] with Jim at the keyboard and Mary to his left. In this module, they used properties of exponential growth and polar coordinates to understand why these spirals that occur so often in nature (such as in the shell of the chambered nautilus) are called equiangular. About eleven minutes into their work, they have finished measuring the outer spiral of a cross-section of a nautilus shell superimposed on a polar graph. They had worked together, making steady progress and had encountered no difficulties. Meanwhile the investigator was in the office with them, but not closely observing their work because they seemed to be progressing well and did not seem to need or want any help. At the point at which this vignette begins, the subjects had just read the following question: "Follow the instructions in your worksheet to plot the sequence of radial measurements, r_n, as a function of the counter n. What sort of growth does this look like?"

The subjects noted that the graph of the data seemed to be exponential. They read the next instruction, "Experiment with logarithmic plotting of the data to determine the type of growth," not paying attention to that (or perhaps not understanding what it was asking). They then went on to the next instruction, "Find a formula for a continuous function $r = R(t)$ such that $R(n)$ reasonably approximates the nth measured radius, r_n." They then tried to guess what the formulas would

[2] http://www.math.duke.edu/education/ccp/materials/mvcalc/equiang/index.html

be by noting that for each increase of 1 in n, r_n seemed to increase by a factor of 1.5. At this point, they could not figure out how to get *Maple* to plot the function 1.5^x. Mary immediately suggested that they check the help menu (?plot in *Maple*). This helped and they were able to make a plot. Mary stated "we have a problem with the range"; Jim concurred and they attempted to fix the problem. At this point (about eight minutes after they began grappling with the problem of plotting a function to match the data) the investigator asked, "What are you trying to do?" After looking at what they were doing, he explained how to more easily use *Maple* to graph a semi-log plot of the data (a data analysis tool that helps find formulas for data that are exponential). The subjects then worked for another three and a half minutes, each concentrating and contributing, while the investigator stepped back. At that time, about twenty-four minutes into the exercise, the investigator asked them "Is that working?" Jim answered, "No. Here's the problem. Hold on. We have to do something first." Jim fixed a *Maple* command and then asked the investigator a few questions to clarify what *Maple* commands will create the plot he needs. Two minutes later the students succeeded in producing a semi-log plot that was approximately straight. They continued to struggle, largely with the software, for another seven minutes before they succeeded in superimposing the graph of the exponential function on the data plot in a way that produced a good fit. They were clear on the mathematical principles, correcting each other when mistakes were made. They then went on to the next part of the module.

Discussion. In this vignette, the students struggled with the software, but were not frustrated and did succeed in accomplishing what was asked. Much of the subjects' effort in this vignette involved getting *Maple* to do the computations that they wanted done. Because Mary and Jim were hesitant to ask for help, they proceeded through the module at a slow pace. This raises certain questions: Was this an efficient use of the student's time? If this was a classroom and the investigator was the instructor, should he have intervened more often to help the students move along more quickly? How does the instructor know when and in what situation to intervene?

We also noticed in this vignette, as in others, that students are willing to persist for a long time trying different things on the computer. In particular, trial and error can be much quicker on a computer than with pencil and paper and because of this, students are less likely to be demonstrate self-regulatory behavior.

This vignette illustrates three issues: (1) as in any active classroom, there are the questions of when and how the teacher should intervene, (2) the computer learning environment seems to affect students' ability to manage their time efficiently and (3) whereas computers solve certain pedagogical problems (such as time consuming calculations), they create others (such as learning the nuances of the software).

Vignette 2. This vignette also involved Jim and Mary and began several minutes after Vignette 1. They were asked to "construct a function $r = r(theta)$ that describes the shell radius as a function of polar angle." They typed in:

$$x := theta \to R * cos(theta); y := theta \to R * sin(theta)$$

Because R was defined earlier as a function (not as a constant), *Maple* would not plot the parametric equations. After the subjects struggled with this for about four minutes, the following conversation took place:

Investigator: Did you run into a problem?

Jim: When we graphed it, we didn't get anything.
Investigator: What is R? You have $x := theta \rightarrow Rcos(theta)$. R is the radius, right?
Mary: Yeah.
Investigator: Isn't that changing as a function of theta?
Jim: Oh. So we have our R formula.
Investigator: You can just say $R(theta)$. Does that make sense?
Jim & Mary: Yeah.

They tried that and it worked. They were then able to proceed through the next parts of the module.

Discussion. In this vignette, the subjects are confronted with a mathematical problem that they may not have been able to solve on their own, and without solving it, they could not proceed through the module. Whereas in the first vignette the investigator/instructor helped move things along, in this vignette it seems that the intervention of the instructor was essential. This was not an isolated instance of this sort; we noticed such problems often as we reviewed the data. This raises the question of what would happen here in the absence of an instructor, e.g., in a distance learning situation. It also raises the issue of whether and how such problems can be anticipated by curriculum developers and what possible technological solutions (such as links to hints) can be built into the module.

Vignette 3. Again we look at Jim and Mary working on the Equiangular Spiral module, this time towards the end of the module. They read the instructions: "Find derivatives of x and y with respect to *theta*, and then combine the results to find dy/dx in terms of *theta*. You may want to use your helper application for this." They could not remember how to get *Maple* to compute derivatives and began trying to find the derivative of $x = r_0 e^{k\theta} \cos\theta$ with respect to θ. After a few minutes of not being able to do this, Mary became frustrated and said, "We can just do it by hand." She began to do the calculation on paper, but Jim said, "I'm trying to remember how *Maple* works." After about a minute, Mary finished and they had the following conversation, which was conducted in a friendly and jocular manner:

Mary: O.K.
Jim: Shut up.
Mary: (laughs) Here, it's just the product rule.
Jim: Yeah. It would be nice if *Maple* would do it for us.
Mary: It will.
Jim: Yeah. I want it to do it.

A minute later, working together they got *Maple* to do the calculation.

Mary: You see, it's exactly what I just did.
Jim: Yeah but your way is stupid.
Mary: But it was quicker.

The instructions asked them to divide $dy/d\theta$ by $dx/d\theta$ to get a formula for dy/dx. This would have been quite tedious with pencil and paper but they were now able to use *Maple* to do this computation in a couple of seconds. The instruction then directed them to evaluate an even more complicated expression that reduced to $1/k$. They got this result with *Maple* (by now, they were using *Maple* correctly), and Jim said, "Wow. I want to work this out on paper. I don't believe that."

Discussion. Note that Mary first suggested using pencil and paper but then wanted to use the CAS, whereas Jim first insisted on using the CAS but in the end did not believe the results without checking using pencil and paper. The main issue raised by this vignette is how and why students choose one tool or another. As we've seen in other videotaped sessions, the students used the CAS and hand-held calculators, as well as pencil and paper. Some students (like Mary and Jim above) seem to believe their pencil and paper calculations more than the results of a CAS application. This situation may have been exaggerated by getting the surprising result of the complicated expression reducing to $1/k$. This issue of what tools to use to solve a particular problem raises several questions: Does familiarity and comfort with the tool affect how readily one accepts the results produced with that tool? Do some students (particularly those who have been successful in school) receive some ritualistic satisfaction from doing pencil and paper calculations? Or, alternatively, is it that in situations like this, where it seems that solutions are pulled out of a hat, that students want to do the hand calculations because of a intrinsic motivation to understand the surprising results? How do students check their work, given multiple technological tools?

Vignette 4. In this vignette, we examine the work of Andy and Larry working on a module called "Correlation and Linear Regression"[3] with Andy at the keyboard and Larry to his left. In this module, they learn about correlation and use that notion to develop an understanding of linear regression and least squares estimates for linear data. At the beginning of the module, the students are given the scores of 15 students on two tests and are asked to plot the scores on test 1 vs. the scores on test 2. The first thing they did was follow a link to another file that showed them how to use *Maple* to make a scatter plot. Andy correctly cut and pasted the zip command that tells *Maple* to create coordinates from two sets of variables. They then needed to decide what the x and y should represent (the correct answer being test1 and test2). The following conversation then occurred. During this time they were in the room by themselves; the instructor had stepped out for a few minutes.

Andy: (types and says out loud): our data = $zip(x, y) \rightarrow [x, y]$
Larry: What does the zip do? Is that just something in the definition?
Andy: I have no idea. It just says it right there.
Larry: You just copied it?
Andy: Yeah. Why not.
Larry: I thought it was a cool function, it sounded interesting. I guess just test 1 test 2
Andy: No. No. We're not planning on plotting them against each other.
Larry: Yeah we were. Weren't we?
Andy: No.
Larry: I thought we were plotting test 1 vs. test 2.
Andy: No. We're plotting test 1 and test 2 so we want to do these against, just like, the one, so each number represents a 1.

[3] `http://www.math.duke.edu/education/modules2/materials/test/test/`

Larry: Oh wait, we're just putting the plots on the same graph and not plotting them against each other?
Andy: Yeah.
Larry: OK.
Andy: I don't know if this is going to work.

Since what Andy said made no sense, they made very little progress. At first, they typed "plot (test 1)" which produced nothing. Andy then spent a couple of minutes poking around in the help menu, thinking that he was having syntax problems rather than a problem understanding the mathematics. After about four minutes, Andy said, "You know what we can do?" and plotted the test1 data vs. the set $\{1,1,1, \ldots, 1\}$. Larry said, "That can't be the normal way to do it. Interesting though." Periodically, throughout this process, Larry politely and without being assertive asked whether Andy was sure that they were not supposed to be plotting test 1 vs. test 2. We should note that the module has a link to the glossary for the word "versus," yet neither Larry nor Andy suggested following that link. Finally, they looked further down on the worksheet and realized that they should have been plotting test 1 vs. test 2. They typed this in and got the correct scatter plot. At this point the investigator entered and verified for them that they were on the right track. Interestingly, Andy didn't seem embarrassed by his refusal to take Larry's advice and Larry didn't seem to blame him.

Discussion. Several issues surface in this vignette. Perhaps foremost among these issues is the ongoing tension between the desire of one student to understand the conceptual ideas embedded in the mathematics problem and the other student's push to solve the problem and to get through the assignment as quickly as possible. For example, Larry asks, "What does the zip do?" Andy indicates that he has no idea what the zip does, seemingly implying that the primary goal is to finish the problem, not necessarily understand the underlying ideas and processes. This raises the question: How can computer learning environments be designed so that they foster learning for understanding but that also use students' time efficiently, avoid student frustration, and prevent those students who have a desire to get the work done with a minimum of time and effort from missing the point of the lesson?

A second issue raised by this vignette is similar to an issue in Vignette 1. Students struggle with the tools (*Maple* software commands) as much as they do with the mathematics concepts. Andy spends much of his time in this vignette trying to correct what he perceives to be a *Maple* syntax problem, when the real problem is his lack of understanding of the concept of "versus." Perhaps with greater metacognitive awareness, Andy might be able to ask himself whether he is having a problem with the tool or the concept? This raises the following question: In a computer learning environment is it possible to build into lessons ways to help students develop metacognitive skills? For example, in this case, why did neither Larry nor Andy suggest following the link to the glossary for the word "versus"? What can instructors and module writers do to strengthen students' abilities to recognize and differentiate between tool problems and conceptual misunderstanding? Can the learning of metacognitive processes be embedded and situated in computer based modules without significant costs in terms of instructional time?

Another issue, which surfaces in this vignette, concerns the role of the teacher. In this particular scenario, the investigator permitted the two students to struggle independently for quite some time, prior to interacting with them. This did allow

Andy and Larry eventually, after a significant expenditure of time, to discover on their own how to plot the data. However, would this be an efficient use of instructional time? Would more learning have occurred if the instructor had intervened sooner? What is the cost-benefit ratio in terms of letting students discover solutions versus intervening and more directly guiding the computer based modules? Again, as was discussed in Vignette 1, these are questions in any active classroom.

A final aspect of Vignette 4 concerns the role of student-to-student dialogue in mathematics computer learning environments. In this particular vignette, Andy and Larry apear to be sharing ideas, but Andy is not seriously considering Larry's questions and Larry is not asserting himself. Andy continues to plod down the wrong path, despite Larry's early suggestion that they should plot test 1 versus test 2. In Vignettes 1, 2, and 3 genuine dialogue between the pair of students seemed to exist. The students shared their provisional hypotheses and provided one another useful feedback. In the case of Andy and Larry, Andy emerges as an assertive but conceptually mistaken leader. Larry, who correctly understands the mathematics problem, remains for the most part a passive follower. This raises several questions about the quality of interactions in any active collaborative learning situation, and in particular in the types of technology rich environments described here: In an interactive computer learning environment where meaningful student dialogue is essential for developing understanding, what steps can the instructor take to facilitate dialogue? How can mechanisms be built into the computer modules (or other active learning materials) to get students to reflect on the quality of their interactions? How does the instructor structure the lesson to minimize the problem of one student taking over the learning situation? Can interdependence and shared responsibility, as well as other aspects of cooperative learning, be more effectively built into computer modules?

Vignette 5. In this vignette, Carl and Kevin are working on the module called "Correlation and Linear Regression" and after examining formulas for the correlation coefficient are asked to compute the correlation coefficient for a data set consisting of four points: $\{(1,2), (2,3), (3,4), (4,3)\}$. They plunged into the exercise using pencil and paper but when (three and a half minutes later) they were faced with trying to calculate the square root of 2, Carl asked Kevin, "Do you have a calculator?" After Kevin looked around and couldn't find one, the investigator said, "What are you looking at?" (referring to the computer). The investigator expected that the subjects would bring up *Maple* but instead Carl brought up the computer's scientific calculator (a standard accessory in MS Windows). Five minutes later they finished the calculation. The following conversation occurred. (Note: computing the standard deviation is a step in using the given formula for the correlation coefficient.)

Investigator:	Would it have been easier or harder to figure out how to get *Maple* to do the standard deviation?
Carl:	If we had a nice little equation like that it would be fairly easy I think.
Investigator:	Actually didn't we do that before?
Carl:	Up here? Oh the coefficient thing. Yeah.
Kevin:	Oh yeah that would have been easier.

Discussion. In almost every videotaped lab session, including this one, the researchers noted three aspects of students' use of tools in computer based learning environments: (1) the perceptions students have of the array of "tools" available to them to use to solve a problem; (2) students' notions of when and where it is appropriate to use a particular tool; and (3) and the degree to which students believe or "trust" that a certain tool is a reliable means of producing the correct solution to a problem. For example, after beginning their work using pencil and paper, Carl and Kevin then used the computer's calculator. By the end of the session, with the guidance of the instructor, Carl and Kevin begin to reflect on their choices of tools and they seem to realize that *Maple* is a powerful tool they also have at their disposal.

This vignette raises several potential research questions. For example, even though Carl and Kevin had been using *Maple* for two or three months, the idea of using the computer algebra system to calculate a standard deviation did not occur to them. This raises the following questions. Prior to putting students into a computer based learning environment, can we design ways to introduce students to computer algebra systems that help them feel comfortable using them? Can we design computer modules in ways that get students to think more reflectively about their choice of tools (pencil/paper, calculators, computer calculator, CAS)? What factors underlie students' perceptions of the accuracy, reliability and efficacy of a particular tool?

Vignette 6. Again, we describe the work of students working on the module "Correlation and Linear Regression." The students, Alex and Neil, are roommates and are both taking linear algebra. After a couple of minutes of work, the following conversation occurs:

Alex: Why don't you type dude?
Neil: Are you sure?
Alex: Yeah, yeah.
Neil: I thought you wanted to type?
Alex: You're better with commands.
Investigator: So what's the deal? Does he usually?
Alex: Uh, he uh, he did before because he knew *Maple* and I took the last couple.
Neil: We take turns.
Alex: Yeah, it's his turn anyway.

After this, they returned to work on the module. During this period, they sometimes thought aloud and sometimes talked to each other (but they looked at the screen rather than each other) and they often pointed at objects on the screen as they talked.

Soon after they switched seats, Neil asked the investigator, who was still overseeing their work, "How do we turn radicals into decimals?" (*Maple* output is exact unless you ask for the decimal approximation.) The instructor/investigator, who was present responded, "you go evalf" (evalf is the command to compute decimal approximation). Several minutes later, after the instructor/investigator left and was sitting in the adjacent room, they acknowledged in their discussion that they did not know how to interpret the word "versus" (as in "test 1 versus test 2"). Though the word "versus" was highlighted as a hot link, they did not click on the link. Eventually, they figured out what versus meant using the context of questions that came up later.

About ten minutes into the tape, when they executed the command to plot the scatter plot of test 1 versus test 2, the plot was displayed incorrectly. This was not the result of any error on the students' part but was due to some technical problems concerning the memory of the machine and *Maple*'s interaction with the hardware. The students' first reaction was to assume that the output was correct and they tried to construct some meaning out of the incorrect output (obviously this was a challenge). After realizing that the output was incorrect, Neil and Alex tried all sorts of things such as changing the format and display options hoping to get the graphs to come out right.

Discussion. One issue that apeared in many of the lab sessions concerned the question which of the lab partners would assume certain roles; for instance, who will take responsibility for typing, using the mouse and offering initial ideas. While in most cases the roles are not explicitly discussed, in this case Neil and Alex directly addressed the issue of who would use the keyboard. This raises the questions of whether roles should be assigned, how students work out roles in the absence of assignments and what impact, if any, do the roles have on learning.

This ties in again to the question of the role of the instructor: Should the instructor assign roles to students? Although assignment of roles has been discussed in the literature on cooperative learning (Slavin, 1995), do different issues arise in the interactive computer environment?

Another issue in this vignette concerns the different ways that students seek help. Neil and Alex encountered difficulty in interpreting "versus," yet they either failed to recognize that versus was a hot link in the HTML document, or they were reluctant to use it. Why do students sometimes use links and other times ignore them? What cognitive conditions prompt students to use hot links? This vignette also re-emphasizes issues seen in earlier vignettes: What are the ways students seek and receive help in the computer learning environment? What is the role of the instructor in providing support and guidance?

A related issue that appears in this vignette has to do with the intellectual dialogue between the students during the session. One of the strongest features of the interactive computer lab approach seems to be the way that the labs foster collaborative discourse. This type of dialogue is seen in this vignette when Neil and Alex think aloud, offer hypotheses, and make predictions. However, the question arises: Is this type of dialogue typical of this computer learning environment? Does this environment foster meaningful collaborative discourse in ways that would not occur without the computer, and why?

A final issue that surfaced in this vignette was the technical problem with the interaction of the software and hardware. The hardware memory problem was very common and frustrating to many students (and the investigator!). This was not consciously built into the study but it points out the serious difficulties that can occur when technical problems happen. Typically, students think they did something wrong, as in the reaction of Carl in a different session: "Oh! What did I change? The computer hates me." Often, even the instructor (as in this case) cannot solve the problem and this can seriously upset the flow of the lesson, especially if it depends on the computer output. This raises the same question we discussed in the Vignette 5 concerning the impact that these technical problems have on the sense of credibility and trust that the students have in the computer as a tool. Another question concerns a metacognitive issue: How can we help students learn how to check the reasonableness of an answer and determine whether discrepancies are due to mathematical or technical errors?

Vignette 7. In this last vignette, we describe the work of Dan and Aaron working on the module "Correlation and Linear Regression." Dan and Aaron are among the best students in a linear algebra class. Dan is an electrical engineering major and Aaron is a mathematics/economics major. Dan was working comfortably at the keyboard while Aaron was thinking critically and actively about the questions being asked. They had just examined the formula for the correlation coefficients and were asked to compute the correlation coefficient for a data set consisting of the points: {(1,2), (2,3), (3,4), (4,3)}.

Aaron: Could you go (scroll) up so I can see the formula?
Dan: There's a way to do this in *Maple*. I don't remember how.
Aaron: To find r?
Dan: To sum a list. I don't remember how right now.
Aaron: You can do the standard deviation, can't you? (meaning on *Maple*).
Dan: Yeah.

Dan enters "$X1 := [1,2,3,4]$; $Y1 := [2,3,4,7]$;" and tries to remember the syntax for standard deviation.

Aaron: It's at the top if you can't remember.
Dan: (mutters) I can't remember.

Dan checked the syntax and computed the means and standard deviations of X1 and Y1 using *Maple*. Aaron began to input that information, using pencil and paper, into the formula for the correlation coefficient. While Aaron got involved in the pencil and paper computation, Dan, intently viewing the computer screen, tried to remember how to sum a list. He used trial and error but did not use the help menu.

Aaron: I can do these by hand (referring to summing the product of x_i and y_i).

Aaron completed the computation, making some errors and got $3/80 * sqrt(10)$. Dan had *Maple* evaluate this and got approximately 0.12.

Aaron: That's really low.
Dan: Yeah, these two are correlated really well. It's got to be higher than that.

Dan tried to create a scatter plot, while Aaron and the investigator tried to find the arithmetic error in his calculation Dan ran into trouble getting *Maple* to plot the scatter plot.

Dan: It doesn't seem to show for some reason.
Investigator: That's strange.
Dan: The graph shows three points, not four.
Investigator: Where would the fourth point be?
Dan: There should be one right here but it (the correlation) still should be better than 0.1.

Aaron continued looking for the computational error during this interchange and the investigator returned to helping him, suggesting that instead of looking for the error, he simply redo the calculation. Meanwhile Dan, quietly and persistently, tried to get *Maple* to compute the correlation coefficient by using the sum command but still couldn't get it to work.

> Dan: I do this all the time in my other class. (He uses *Maple* in his electrical engineering class.)
>
> Investigator: We'll check it with *Maple* later.
>
> Aaron: I think it's $2/sqrt(10)$. I think that's right. I made two algebra mistakes and I found both of them.
>
> Dan: (evaluating the expression on *Maple*) Yeah, that's reasonable.

In writing up the answer and explanation, Aaron made suggestions. Dan listened to Aaron's suggestions while also incorporating his own ideas (though not saying anything), using Aaron's input. After reading what Dan typed, Aaron said, "I think that explains it."

Discussion. In this vignette we see several of the issues we observed in other vignettes. For example, the two students struggle with computer software. In this case they can't remember the *Maple* syntax needed to sum a list. This raises the question: How do we provide students with adequate training in computer algebra software without taking away from instructional time? A second issue raised in this vignette is the problem students apear to encounter with self-regulation and time management. Dan inefficiently used trial and error to discover how to sum a list. They devoted too much of the lab time trying to create a scatter plot. This raises several questions: Are students more likely to have difficulty managing their time and regulating their thought processes in a computer based mathematics lab? As anyone who uses a computer knows, sometimes a task which at first would apear to take a few minutes on a computer can end up taking much longer, perhaps because it is so easy to explore options on the computer.

This vignette also raises the question of the role of the instructor in an interactive computer environment. How directive should the instructor be? In this case, the instructor finally intervened and directed Aaron to repeat the calculation, instead of inefficiently looking for his error. What responsibility did the instructor have in this vignette for making sure the lesson progressed at a reasonable pace?

Aaron and Dan engaged in productive collaborative discourse and cooperative problem solving. There are several good examples of the in-depth cognitive processing that conceptually based approaches to mathematics are designed to elicit. For example, when Aaron and Dan compute a correlation coefficient of 0.12, they experience a sense of cognitive disequilibrium. "It's got to be higher than that," Dan remarked. The two students appear to have made a prediction or estimate of higher correlation coefficient based on their initial analysis of the data. When they computed a lower number, they began to question its reasonableness. Is questioning the reasonableness of an answer more or less likely to occur in a computer-based program? Will some students tend to accept as reasonable what the computer produces, because they view the computer as infallible? How can we design lab experiences which encourage students to make predictions, estimate reasonable answers, and then compare their estimates to computer generated products? How can we build into computer lesson modules tasks which elicit the high level mathematical cognitive processing we want students to engage in?

Summary and Conclusions

The research presented here is the first stage of a process whose purpose is to develop an understanding of how students learn in technology-rich environments for the mathematics classroom. By carefully observing student work in a particular technology-rich environment (i.e., the Connected Curriculum Project), we generated a set of questions for further study. These questions were not derived a priori from a theoretical perspective but were derived from the data. Using grounded theory as a methodological approach, the data served as a first step in constructing a theory that will explain learning in this environment.

The analysis of the data led to the formulation of three categories of research questions: (1) What is the role of the instructor in this environment? (2) What types of behavior and thinking processes are students engaged in as they work together in front of the computer, and how can the modules be written to facilitate students' self-monitoring and effective collaborative interaction? and (3) What opportunities and obstacles are raised by the technology itself? Research in each of these areas has important implications for curriculum developers, mathematics instructors, and students.

For each of these categories, we will summarize some of the issues and questions that arose from our observations:

1. *The role of the instructor.* As in any active classroom, the instructor must confront questions of when and how to provide support and guidance to students who are engaged in complex problem solving. We observed incidents where the teacher's intervention was critical for the student's progress. In other cases, we observed students floundering and making little progress because of the lack of help. We also saw students struggle, perhaps inefficiently, with a problem, but eventually solve it on their own. The questions raised here include: What should be the role of the instructor in a computer based mathematics class? How does the instructor make decisions about intervening? In an interactive computer mathematics lab, how does the instructor ensure that instructional time is used efficiently? Is learning more likely to occur if the instructor intervenes whenever students encounter difficulties? Should the instructor allow students ample time to discover solutions on their own? How, if the course is being delivered via the Internet, could software developers simulate the role of the instructor in the lab? For example, what kinds of hints (or nested sequence of hints) could be provided and how can this be done intelligently (in the sense of artificial intelligence)?
2. *Types of behavior, thinking processes and self-monitoring as students engage in collaborative interaction.* Much of the focus of our observations was on the ways students' chose to work with one another as they engaged in collaborative problem solving. For example, we observed students making decisions about which tools to use to solve the problems presented to them on the computer. One of the recurring themes concerned how, when, and why students chose one tool or another (e.g., pencil and paper, a calculator and/or CAS). Analysis of the data raised many questions

concerning the perceptions students have of the array of tools available to them. We also observed students struggling with what they thought were tool problems, when in fact they had conceptual misunderstandings. How can students recognize and differentiate between tool problems and conceptual misunderstanding, and how can instructors help them?

We also observed how, as students worked in different situations, lab partners would assume certain roles such as hypothesizer, verifier, and recorder. We also saw students deciding who would take responsibility for typing, using the mouse, and offering initial ideas. At times, decisions of this sort were consciously made; at other times, the students seemed to choose roles without discussion. These observations raise the question: would students benefit from assigned roles? Should roles be structured to minimize the problem of one student taking over the learning situation?

We saw some students trying to understand the underlying concepts while others were trying to get through the lab as quickly as possible. In an interactive computer learning environment where meaningful dialogue is essential for developing understanding, can mechanisms be built into the computer modules to get students to reflect on the quality of their interactions? Can interdependence and shared responsibility, as well as other aspects of cooperative learning, be built into computer modules?

We also noticed students using (or not using) links in many different ways. Some of these hot links were labeled as hints; others were links to a glossary or other resources. Some clicked on the hint hot links immediately, some waited and clicked if necessary and some didn't link to the hint even when they were stuck. What cognitive conditions prompt students to use hot links? What are the ways students seek and receive help in the computer learning environment? How can we get students to use help tools built into computer environments so they spend less time floundering and getting frustrated on syntactical problems? These questions about links are questions of both cognition (how students are thinking about the problems at hand) and metacognition (their thoughts about their thinking).

We saw instances of productive dialogue as students tried to solve the problems presented to them. What produces this type of constructive dialogue: the computer learning environments, the content presented, the group environment or project-oriented nature of the work? In other words, how do the physical, intellectual, social and pedagogical environments interact to produce learning? Does the computer lab environment foster meaningful collaborative discourse in ways that would not occur without the computer, and, if so, why? What steps can be taken to facilitate dialogue in computer-based mathematics classes?

In terms of self-regulation, the computer learning environment seemed to affect the students' ability to manage their time efficiently. Is there a positive side of students becoming so involved in problem solving that they lose track of time? Where should the balance be between open-ended discovery versus a focus on getting the correct answer in a set amount of time? We have noticed that students, because they must start doing something even before they've understood the problem, often do not reflect before doing a calculation. In fact, this is true of students using hand calculators or pencil and paper as well as students using computer algebra systems. Are students more likely to jump into difficult computations or use guess and check strategies when they have access to a computer algebra system than when they are using hand calculators or pencil and paper? To what extent does this strategy promote learning even when they are going down the wrong path entirely? Can the learning of time management, self-regulation, and metacognitive processes be embedded and situated in computer based modules without significant costs in terms of instructional time? And (relating to problems with the technology), how can we help students learn how to check the reasonableness of an answer and determine whether discrepancies are due to mathematical or technological errors?

3. *The technology itself.* Whereas computers solve certain pedagogical problems (such as time consuming calculations), they create others (such as learning the nuances of the software). Students struggled with the tools (*Maple* software commands) as much as they did with the mathematics concepts. And the technical problems that occurred with the interaction of the software and hardware raise questions of how students and teachers react to such problems.

   Clearly not all these questions fall neatly into a single category. For example the question of whether the instructor should assign roles to students depends on understanding what roles students assume, how they assume those roles, and the effect those roles have on learning. This is not an exhaustive list of issues. We can imagine many other questions, but, consistent with the principles of grounded theory, we are limiting the discussion to the evidence presented in the data.

   In interpreting these data, it is important to realize that these students were talented students doing mathematics at a level beyond calculus and using specific software in a laboratory setting. It is not our purpose here to generalize these results to a larger population but to use these observations to suggest areas for future study. It is also important to note that each entering class of students brings more familiarity, more comfort and more sophistication with using educational technology. It is not clear which problems faced by the subjects in this study will likely be problems for students several years from now.

There were many issues not addressed in this study such as method of instruction and assessment, the physical environment of classroom and affective issues. These are certainly areas for future study. And as the categories above are reformulated and refined as we collect more data and develop a theory, the need to triangulate our observations (that is to verify our observations by interviewing subjects and collecting other sources of data) will be crucial.

Above, we have raised many questions that are suitable for immediate and more focused study. In conclusion, we suggest several follow-up studies based on the following questions:

1. What are the implications of these observations for distance education? What would be different when students work alone as opposed to working

with a partner? Will they learn the "right" lessons? How (or can) problems such as the ones identified in this study be anticipated by curriculum developers of distance learning materials and what possible technological solutions can be built into these materials?
2. Which of these issues arise in a real classroom setting versus an experimental setting? Which don't? What other issues arise?
3. Initially, we thought the person who had control of the keyboard might have been the more active learner, but we have seen instances where the person who was not burdened with keyboarding was free to think more about the mathematical content. We would suggest follow-up studies that focus on particular aspects of the way students work together. These studies should include clinical observations along with interviews and other data collection.
4. How does the speed, allure and stimulation of computers affect the ways students solve problems? We would suggest focusing in great detail (perhaps with think aloud protocols) on a comparison of students working on a task with pencil and paper as opposed to computers. Among other things, such a study could document how computers affect student time management.
5. What can we learn from existing research in other areas (e.g., cooperative learning, problem solving approaches to instruction)?
6. How are the questions raised here different in an active learning environment without computer technology?

We find this last question to be particularly important. Many of the questions and issues raised by the current study on interactive technology rich learning are relevant to active learning environments in general. For example, the question of how an instructor motivates students to reflect on the quality of social interactions in a collaborative work environments is important in all cooperative learning situations, not just technology-rich paired learning activities. Throughout our work on this research project, we found ourselves asking whether a particular instance of behavior was unique to a computer-based learning environment or more relevant to all active learning situations. As Dubinsky and other researchers (Asiala & Dubinsky, 2000; Dubinsky & Schwingendorf, 1997) have pointed out, it is difficult to differentiate the impacts of technology from the impact active or cooperative learning on student understanding. This is one of the first questions we plan to address in future research.

As we acknowledged, many of the important and interesting questions concerning cooperative learning in interactive technological environments have been examined by previous researchers. In our grounded approach, many of these same issues emerge. Another important study that needs to be done is a comprehensive review of the research literature that documents what is known about each of these questions and what remains to be learned.

Our long-term research program involves the creation and validation of a model of learning and teaching of mathematics in a technology rich environment. The model will examine the nature and importance of the relationships among the following components:

student
teacher

content and context
materials (software, text)
method of instruction and assessment
physical environment of classroom
affective environment (e.g., classroom atmosphere)

We plan to identify sub-components of these factors. For example, students' issues might include student's prerequisite knowledge, attitudes, motivation, and learning style. We hypothesize that there will be significant interactions among the sub-components both within and between larger components of the model. We realize that this is a very bold and ambitious agenda. We plan, working with others over a period of years, to make progress toward an understanding of the relationships among these components.

References

Asiala, M., & Dubinsky, E. (2000). *Evaluation of research: based on innovative pedagogy in several math courses.* Unpublished report.

Battista, Michael T. (1999). The mathematical miseducation of America's youth. *Phi Delta Kappan, 80*(6), 425–433.

Bookman, J. & Blake, L. D. (1996). Seven years of Project CALC at Duke University—Approaching a steady state? *PRIMUS, 6*(3), 221–234.

Bruner, J. S. (1966). *Toward a theory of instruction.* New York: Norton.

Chambers, Jack & Bailey, Clare. (1996). Interactive learning and technology in the U. S. science and mathematics reform movement. *British Journal of Educational Technology 27*, 123–133.

Colvin, M. R., Moore, L., Mueller, W., Smith, D., & Wattenberg, F. (1999). Design, development, and use of web-based interactive instructional materials. In G. Goodell (Ed.), *Proceedings of the Tenth Annual International Conference on technology in collegiate mathematics.* Reading, PA: Addison–Wesley.

Cooper, Marie A. Cautions and considerations: Thoughts on the implementation and evaluation of innovation in science education. In A. Kelly, & R. Lesh (Eds.), *The handbook of research design in mathematics and science education* (pp. 859–876). Mahwah, NJ: Lawrence Erlbaum.

Coyle, L., Moore, L., Mueller, W., & Smith, D. (1998). Web-based learning materials: Design, usage, and resources. *Proceedings of the International Conference on the teaching of mathematics.* (pp. 71–73). Somerset, NJ: Wiley.

Davidson, N. (1990). *Cooperative learning in mathematics: A handbook for teachers.* Menlo Park, CA: Addison-Wesley.

Dubinsky, E., & Fenton, W. (1996). *Introduction to discrete mathematics with ISETL.* New York, NY: Springer-Verlag.

Dubinsky, E., & Schwingendorf, K. (1997). Constructing calculus concepts: Cooperation in a computer laboratory. In E. Dubinsky, D. Mathews, & B. Reynolds (Eds.), *Readings in cooperative learning for undergraduate mathematics* (MAA Notes no. 44, pp. 225–246). Washington, DC: Mathematical Association of America.

Fosnot, C. T. (1993). In J. G. Brooks & M. G. Brooks (Eds.) *In search of understanding the case for constructivist classrooms* (p. vii). Alexandria, VA: Association for Curriculum and Development.

Gall, M. D., Borg, W. R., & Gall, J. P. (1996). *Educational research: An introduction.* White Plains, NY: Longman.

Glaser, B. G. & Strauss, A. L. (1967). *The discovery of grounded theory.* Chicago: Aldine Publishing.

Heid, M. K. (1988). Resequencing skills and concepts in applied calculus using the computer as a tool. *Journal for Research in Mathematics Education, 19*(1), 3–25.

Heid, M. K., Blume, G., Flanagan, K., Iseri, L., Deckert, W., Piez, C. (1998). Research on mathematics learning in CAS environments. In G. Goodell (Ed.), *Proceedings of the Eleventh Annual International Conference on technology in collegiate mathematics* (pp. 156–160). Reading: Addison-Wesley.

Krantz, Stephen. (2000). Imminent danger: From a distance. *Notices of the AMS 47*(5), 533.

National Council of Teachers of Mathematics. (1991). *Professional standards for teaching mathematics*. Reston, VA: Author.

National Research Council. (1991). *Moving beyond myths: Revitalizing undergraduate mathematics*. Washington, DC: National Academy Press.

National Science Foundation. (1996). *Shaping the future: New expectations for undergraduates for undergraduate education in science, mathematics, engineering, and technology*. Washington, DC: Author.

Papert, S. (1980). *Mindstorms: Children, computers, and powerful ideas*. New York, NY: Basic Books.

Pea, R. (1987). Cognitive technologies for mathematics education. In A. Schoenfeld (Ed.), *Cognitive science and mathematics education*. Hillsdale, NJ: Lawrence Erlbaum Associates.

Piaget, J. (1952). *The origins of intelligence in children*. New York: International Universities Press.

Portela, J. (1999). Communicating mathematics through the Internet: A case study. *Educational Media International, 36*(1), 48–67.

Rogers, E., Reynolds, B., Davidson, N., & Thomas, A. (2001). *Cooperative learning in undergraduate mathematics: Issues that matter and strategies that work* (MAA Notes no. 55). Washington, DC: Mathematical Association of America.

Romberg, Thomas A. (1992). Perspectives on scholarship and research methods. In D. A. Grouws (Ed.), *Handbook of research on mathematics teaching and learning* (pp. 49–64). New York: Macmillan.

Slavin, R E. (1995). *Cooperative learning: Theory, research, and practice*. Boston: Allyn & Bacon.

Smith, D. A. (2000). Renewal in collegiate mathematics education: Learning from research. In S. L. Ganter (Ed.), *Calculus renewal: Issues for undergraduate mathematics education in the next decade* (pp. 23–40). New York, NY: Kluwer Academic/Plenum Publishers.

Smith, D. A. & Moore, L. C. (1990). Project CALC: In T. W. Tucker (Ed.), *Priming the calculus pump: Innovations and resources* (MAA Notes no. 17, pp. 51–74). Washington, DC: Mathematical Association of America.

Smith, D. A. & Moore, L. C. (1991). Project CALC: An integrated lab course. In C. Leinbach et al. (Eds.), *The laboratory approach to teaching calculus* (MAA Notes no. 20, pp. 81–92). Washington, DC: Mathematical Association of America.

Vygotsky, L. S. (1978). *Mind in society*. Cambridge, MA: Harvard University Press.

DEPARTMENT OF MATHEMATICS, DUKE UNIVERSITY, DURHAM, NC 27708
E-mail address: `bookman@math.duke.edu`

PROGRAM IN EDUCATION, DUKE UNIVERSITY, DURHAM, NC 27708
E-mail address: `dmalone@duke.edu`

Research in Collegiate Mathematics Education

Editorial Policy

The papers published in these volumes will serve both pure and applied purposes, contributing to the field of research in undergraduate mathematics education and informing the direct improvement of undergraduate mathematics instruction. The dual purposes imply dual but overlapping audiences and articles will vary in their relationship to these purposes. The best papers, however, will interest both audiences and serve both purposes.

Content. We invite papers reporting on research that addresses any and all aspects of undergraduate mathematics education. Research may focus on learning within particular mathematical domains. It may be concerned with more general cognitive processes such as problem solving, skill acquisition, conceptual development, mathematical creativity, cognitive styles, etc. Research reports may deal with issues associated with variations in teaching methods, classroom or laboratory contexts, or discourse patterns. More broadly, research may be concerned with institutional arrangements intended to support learning and teaching, e.g. curriculum design, assessment practices, or strategies for faculty development.

Method. We expect and encourage a broad spectrum of research methods ranging from traditional statistically-oriented studies of populations, or even surveys, to close studies of individuals, both short and long term. Empirical studies may well be supplemented by historical, ethnographic, or theoretical analyses focusing directly on the educational matter at hand. Theoretical analyses may illuminate or otherwise organize empirically based work by the author or that of others, or perhaps give specific direction to future work. In all cases, we expect that published work will acknowledge and build upon that of others—not necessarily to agree with or accept others' work, but to take that work into account as part of the process of building the integrated body of reliable knowledge, perspective and method that constitutes the field of research in undergraduate mathematics education.

Review Procedures. All papers, including invited submissions, will be evaluated by a minimum of three referees, one of whom will be a volume editor. Papers will be judged on the basis of their originality, intellectual quality, readability by a diverse audience, and the extent to which they serve the pure and applied purposes identified earlier.

Submissions. Papers of any reasonable length will be considered, but the likelihood of acceptance will be smaller for very large manuscripts.

Manuscripts should have citations and bibliographies according to the format of the American Psychological Association as described in the fifth edition of the *Publication Manual of the American Psychological Association.*

Note that the *RCME* volumes are produced for electronic submission to the AMS. Accepted manuscripts should be prepared using AMS-LaTeXand the CBMS author packages available from the AMS Web site. Illustrations should also be prepared in a form suitable for electronic submission (namely, encapsulated postscript files).

For further information see the Issues in Mathematics Education section of the Conference Board of the Mathematical Sciences Web site www.cbmsweb.org.

Correspondence. Before submitting a manuscript, send an abstract to the managing editor of *RCME*:

Cathy Kessel
cbkessel@alumni.uchicago.edu

Subsequent correspondence may be with the managing editor, or with the volume editor who has been assigned primary responsibility for decisions regarding the manuscript.

Titles in This Series

12 **Annie Selden, Ed Dubinsky, Guershon Harel, and Fernando Hitt, Editors,** Research in collegiate mathematics education. V, 2003

11 **Conference Board of the Mathematical Sciences,** The mathematical education of teachers, 2001

10 **Solomon Friedberg et al.,** Teaching mathematics in colleges and universities: Case studies for today's classroom. Available in student and faculty editions, 2001

9 **Robert Reys and Jeremy Kilpatrick, Editors,** One field, many paths: U. S. doctoral programs in mathematics education, 2001

8 **Ed Dubinsky, Alan H. Schoenfeld, and Jim Kaput, Editors,** Research in collegiate mathematics education. IV, 2001

7 **Alan H. Schoenfeld, Jim Kaput, and Ed Dubinsky, Editors,** Research in collegiate mathematics education. III, 1998

6 **Jim Kaput, Alan H. Schoenfeld, and Ed Dubinsky, Editors,** Research in collegiate mathematics education. II, 1996

5 **Naomi D. Fisher, Harvey B. Keynes, and Philip D. Wagreich, Editors,** Changing the culture: Mathematics education in the research community, 1995

4 **Ed Dubinsky, Alan H. Schoenfeld, and Jim Kaput, Editors,** Research in collegiate mathematics education. I, 1994

3 **Naomi D. Fisher, Harvey B. Keynes, and Philip D. Wagreich, Editors,** Mathematicians and education reform 1990–1991, 1993

2 **Naomi D. Fisher, Harvey B. Keynes, and Philip D. Wagreich, Editors,** Mathematicians and education reform 1989–1990, 1991

1 **Naomi D. Fisher, Harvey B. Keynes, and Philip D. Wagreich, Editors,** Mathematicians and education reform: Proceedings of the July 6–8, 1988 workshop, 1990